The Future of Surfing in the Anthropocene

Leon Mach
Editor

The Future of Surfing in the Anthropocene

Technology, Environment and Society

Springer

Editor
Leon Mach
School for Field Studies
Center for Tropical Island Biodiversity
Studies
Bocas del Toro, Panama

ISBN 978-3-032-09376-9 ISBN 978-3-032-09377-6 (eBook)
https://doi.org/10.1007/978-3-032-09377-6

This Springer imprint is published by the registered company Springer Nature Switzerland AG
The registered company address is: Gewerbestrasse 11, 6330 Cham, Switzerland

Acknowledgements

I would first like to thank the School for Field Studies for providing the creative space to work on a project like this as well as SENACYT's National System of Investigators program for funding and support. For the inception of the idea to examine the future of surfing in the Anthropocene, I remain indebted to the late Dr. Bob Warren from the University of Delaware. He passionately agreed to supervise my dissertation committee at a time when it was hard to find a faculty member who saw merit in researching surfing, technology and tourism. The time he gave to reading and discussing my work, still influences me, and the desire to grapple with the future of surfing. I would also like to thank my fiancé, Jenn Parker, for her encouragement to pursue this project and her time spent editing and discussing the book. And perhaps more importantly, for helping keep my passion for surfing alive. Lastly, I would like to thank all the contributors to this volume. When I put out the call, I wasn't even sure anyone would answer it. I am deeply honored and grateful for the enthusiastic support for this project and for the incredible group of researchers and practitioners who have generously shared their insights. This book is literally nothing but an idea without them.

Introduction

Although it may seem unconventional, I would like to begin this book by first problematizing the very title of this volume. Following that, I will provide a rationale for how it is organized, along with a preview of each chapter.

Some may find it bold or overly ambitious to take on packed themes like "the future" and "the Anthropocene"—especially in a volume about surfing.

Whether right or wrong, I would not have bothered putting together this volume if I didn't believe that this discussion could provide significant insights into not only surfing but also our contemporary world. It's a world where we, as humans, for better or worse, can make decisions that shape previously unimaginable phenomena, like the composition of the atmosphere and where, when, and how waves break.

The origin of this project dates back to 2022, when a lively discussion about creating waves took place on a surf research list serve I follow. The thread discussed a private company's proposal to create "new surfing waves" in the ocean. The plans suggested "sculpting the seafloor" to engineer a high-performance surf break in Fiji. The stated goal was to turn an area with a poor wave, also referred to as an "almost wave" (i.e., meaning it receives consistently favorable swell and winds, but the bathymetry doesn't quite produce surfing conditions) into a spot with potentially world-class waves by making "minor modifications to the depth, contours, and angle of the reef" (Kushner, 2023).

With seed funding from Scot Nolan (formerly of Space X) and Peter Thiel's (former CEO of PayPal) Founder's Fund, the developers were hoping to demonstrate this theory with a pilot and then profitably replicate it in other areas around the world. They argued that moving wave creation into the ocean would reduce the cost tenfold compared to surf pools, while at the same time having a net positive impact on coastal habitats. While surf pools are growing in popularity, both with surfers and investors, they require the "sea floor" to be created and capable of withstanding constant wave action as well as continuous energy and water inputs to produce rideable waves. On the other hand, since the ocean naturally has a sea floor and water and produces swell, it arguably makes more sense to just tweak the bottom a little bit to create the conditions for ideal surfing. Right? Hell, we are mining the seafloor before we even know what's down there, so why not let surfers manipulate the bottom so we can surf in more places?

I cannot deny a long personal history of looking at these "almost waves" everywhere I have ever lived and imagining what it would be like if they just started perfectly breaking one day. In fact, I was just mind surfing an "almost" right point break yesterday afternoon at Bluff Beach in Panama. But prior to this Fiji proposal, it never occurred to me that seafloor sculpting might become a commercial enterprise. I knew the technology for creating world-class surf in pools which was developing rapidly (as authors in this volume will later discuss), but for some reason, I thought the actual ocean might be seen as out of bounds for this type of anthropogenic manipulation.

Many economic studies, including some I participated in, suggest that surf resources provide significant economic value to coastal communities (Manero & Mach, 2023) and have the potential to revitalize economically depressed regions (Ponting, Teixeira, Costa, Vasconcelos, & Mach, 2023). Given this evidence, it should not have surprised me, or anyone else, that concepts such as wave creation in the ocean might soon become a reality.

Because the technology is out there, and we could theoretically do it, is that reason enough to sculpt the seafloor to create new waves? Perhaps for some, it is, and it very well may be in certain geographical contexts, but I think it's time serious debate occurs around which technologies improve the wellbeing of the collective surf community and the environments where the activity takes place.

In some ways, it seems logical to create waves if we are constantly losing them to climate change and coastal development (Sadrpour & Reineman, 2023). I cannot deny, however, that something makes me feel uneasy about the hubris involved with thinking surfers and developers ought to decide where waves break and that the capitalist economic system will ensure this is done justly with ecologically sound principles that, at the very least, do not make our struggling oceans even worse off. Legendary oceanic explorer Jacques Cousteau's son, Jean Michel Cousteau, echoed this sentiment stating, "My lifetime of experience and moral sense of duty compels me to strongly object to the World Wave Project's proposal to 'sculpt and modify' any reefs to enhance surf breaks. Coral reefs must be protected not destroyed" (Kushner, 2023).

Local Fijian fisherfolk agreed with Jean Michel Cousteau and shot down the project. They decided that the precautionary principle should be taken, and they were not willing to risk the potential damage caused by "sculpting" the reefs, which are ecologically connected to the fisheries they depend upon. So, while the public deliberations may have closed the book on this particular project in Northeastern Fiji, I doubt it will be the last we hear of wave creation in the ocean. Surfers' insatiable desire to ride waves and the Promethean ability to create them will likely find less risk-averse and/or ecologically reliant geographies to experiment. It all seems logical as a technological solution to the problems that are arising with a growing surf population and dwindling surf resources. This is just one of the many provocative examples regarding the nexus of the Anthropocene and the future of surfing.

In a previous volume edited by Jess Ponting and Greg Borne on Sustainable Surfing, I staked the claim formally that we have entered the technological era of surfing (Mach, 2017). The argument suggested that the most significant differentiating characteristic of the modern surf era is the common acceptance of new technological inputs and the feverous pace at which the surf industry accelerates and extends the technological arsenal available to surfers. I broke this down to conceptualize four technological dimensions: (1) physical (e.g., equipment, cameras, wetsuits, transportation), (2) surf forecasting (i.e., advanced understanding of surf conditions and more accurate predictive capacity in more places), (3) internet communication technologies (i.e., the ability to share information and images more easily and from peer to peer, (4) artificial waves (e.g., surf pools, parks, artificial surf reefs, and seafloor sculpting). Despite this articulation and the potential ability for artificial intelligence (AI) to supercharge each of these dimensions, discussion is lacking on the type of future this is leading surfing toward.

This book is meant to stimulate conversations about the future of surfing in the technological era. It will do so by putting out the views of experts from different areas (e.g., government agencies, nonprofit organizations, and academic institutions) in the fields they know best. Some chapters will be academic, and others will be more colloquial, but all will explore an interesting topic and expound on what the future might or could look like in relation to their field.

I must admit, after spending more than a decade researching and writing about environmental governance, I am one of the least optimistic people that I know. Almost every empirical environmental indicator seems to be going in the wrong direction. On top of that, many countries are electing leaders who malign scientific research and deny the existence of a global environmental crisis. Despite this, I do not believe that any future is predetermined.

Reflecting on what type of world we ought to strive for has always been the cornerstone of what guides social movements for change. And while some examinations have found surfers to be politically agnostic (Hill & Abbott, 2009; Laderman, 2014; 2017), other evidence suggests that as waves become threatened by pollution, sea-level rise (SLR), and coastal development around the world, a sea change is occurring in surf culture.

People's identities are now more wrapped up with their hobbies (like surfing) and how they spend their leisure time (a lot of it in the water) than with their work (Bell, 1976; Stranger, 2011). Tens of millions of surfers organize their lives to have access to surfing consistently and surfing has been found to be an important contributor to mental and physical health, as well as, one's general wellbeing, which requires both (Britton et al., 2020; Gibbs et al., 2022; Mach et al., 2024; Wheaton et al., 2021). This is something I know about firsthand.

I decided to research surf tourism because it was a phenomenon that fascinated me. I couldn't help but notice the impact surf tourism was having on many of the towns I just happened to visit as a clueless surf tourist. I started traveling, mostly to Central America, to take breaks from studying and working on environmental, conservation, and development issues, but I quickly noticed that surf tourism was a driver of so much rapid change in places I loved. Surf trips could no longer serve as a break from reality, too much was being developed, and too little was being conserved.

Playa Venao, on the Pacific Coast of Panama, is one example of a place where I camped on the beach with a good friend for a week back in 2008. At that time, there was only one thatch-roofed restaurant on the beach serving mostly whole-fried local snapper, beer, and *patacones*. We had multiple sessions there completely alone, surfing any A-frame peak we wanted to in that huge bay.

The dunes where we camped and were attacked by migratory crabs while sleeping in our board bags, however, are now gone. Booking.com lists more than 80 hotels in the area, and there are many surf schools and places to rent every kind of surf craft. There are also multiple private real estate developments populated by lifestyle migrants (seasonal and permanent) and Airbnbs. I have learned over my years of research that this rapid development is neither good nor bad in-and-of-itself. Sure, I would prefer to be able to keep traveling back there and score with just my crew, but is that really what would be best for the people living in that once remote zone looking for better work and educational opportunities? Understanding the future these communities want and mobilizing them to steer surf tourism development in desirable ways are what has emerged as the best any of us can do in the face of the pressures surf tourism demand brings. Organizing communities and sparking debates is essential for sustainable change. And it is my hope that this volume will contribute to fostering a new wave of dialogue through chapters discussing citizen surf science as well as technologies for conservation and coastal access.

I now live and work in Bocas del Toro, Panama, where I can surf often, research surfing, and continue to teach and write. Ironically, many surf entrepreneurs here seem to channel a Balinese aesthetic when naming and decorating their establishments. A few bars and hotels have even shipped entire wooden facades from Bali to adorn their interiors. While Bali is often held up as the model spiritual, stylish and surf saturated tourism destination to emulate, I can't help but wonder what other aspects of Bali's experience—beyond the architecture and ambiance—might also be unintentionally recreated.

I have experienced and commented considerably on how otherwise remote fishing villages can morph into peri-urban surf towns (some of which have been called surf slums or ghettos), fuelled by investment in making surf tourism a profitable enterprise after being "discovered" by the surfverse (Barilotti, 2002). Surf commentators have suggested that different areas in Bali (where surf tourism began the impetus for visitation) have been showing precursor signs of overtourism since the early 2000s (Barilotti, 2002). Fodor's recently even placed Bali on its 2025 "No Go" list, citing, "Rapid, unchecked development spurred by overtourism is encroaching on Bali's natural habitats, eroding its environmental and cultural heritage, and creating a "plastic apocalypse" (Fodor's, 2024). This is not to say that surf tourism is solely responsible for the current state of tourism in Bali, but it would be equally irresponsible to suggest that it has not made major contributions to both the appealing image of the islands and the volume of foreign visitors to certain areas (Laderman, 2014).

I have worked with towns, like El Zonte and Las Flores in El Salvador trying to control surf tourism growth and nonprofits like Waves for Development, in Lobitos,

Peru, trying to ensure that surf tourism benefits residents and is not overly extractive. Throughout my research, however, the most glaring gap I keep coming back to is that hardly anyone discusses the role of technology in how surfing is progressing as a growing global phenomenon that sparks a lot of travel and lifestyle decisions for an increasing portion of the population.

This book asks how the most stripped-down and ecologically connected sport (Lawler, 2011) became so hyper-technical? Ancient Hawaiians replanted multiple trees for everyone they cut down to make a surfboard and prayed for surf from the gods. Now, we have access to more than one thousand cameras pointed at surf breaks on every continent, smart watches where we can receive texts while surfing, and boards and wetsuits made from petro-chemicals. We surf with drones overhead and cameras in our mouths and can book a surf trip to almost anywhere in the world right from our phones. As Dylan's adage goes, "the times they are-a-changing."

How can those of us, amid this rapid change make sense of it, and what can we do about it?

Why This Project Might Be Doomed (A Disclaimer)

One kind reviewer noted that "any book that purports to discuss the future will inevitably be dated the minute it is published." While this initially stung, it helped me recognize a common reason why people often avoid engaging in the future: the fear of being wrong, or of becoming irrelevant with time. But I've come to see that reluctance as part of the broader problem. When we avoid imagining or debating possible futures, we risk treating the present as fixed and inevitable. This mindset aligns with a form of technological determinism, where speed and efficiency are pursued as ends in themselves, shaping socio-environmental relations in ways that often go unquestioned (Ellul, 1964). In this view, technological development accelerates by feeding off the very problems it generates—for human life, for society, and for the biosphere—while a misplaced faith in technological fixes distracts from deeper, systemic challenges.

The problem is that technological solutions tend to mitigate (at best), deflect, or further complicate rather than resolve much of anything (Norgaard, 1994). As humanity finds itself in an epoch of rapid technological evolution, the pace of innovations and social change can become too dizzying to truly step back and consider where things are going and why (Noble, 1983). When we adapt to the world technological advance creates for us, rather than steer technological advance toward meeting social, psychological, and environmental goals set by society, we end up letting the process and artifacts of technological advance determine the future.

Despite the difficulty, the key is to try and view technology in its present tense (Noble, 1983). How can we understand the current technological ensemble we are enshrined in, what are its component parts, and which aspects improve or detract from the type of lives and environments we wish to have? Only when discussing this can we begin to steer, rather than be steered by technology.

So yes, no one writing in the volume can or pretends to be able to predict the future. Some things we say will likely be dated by the time this volume goes to print. But pondering the impacts our current technological ensemble has upon us and where it all might be heading is still vital. Using surfing as a lens might even help focus the discussion enough to have some sense be made of it all. But that is something for you, the reader, to decide.

The Anthropocene

It is also not lost on me that this volume is being put together at a time when the value of the term Anthropocene is being brought into question. Humans have always altered ecosystems, so how would scientists be able to determine a geological marker for when the earth became so overwhelmingly altered at the planetary scale to be labelled as a new period?

The International Union of Geological Sciences (IUGS) is the scientific body dedicated to deciding how we classify and understand geological eras, past and present. The body has ruled that the earth transitioned into the Pleistocene around 2.6 million years ago and from the Pleistocene into the Holocene 11,700 years ago. They voted in February 2024, however, and the result was that we have not been in the Holocene long enough to transition out of it. This volume accepts that humans are both part of and also heavily influential to all natural and ecological systems regardless of the epoch in which we are currently living. This book will endeavour to look at how the world, altered by human actions, influences the current surf era and what decisions humans are making to adapt to or change the current course.

Looking at the future of surfing in the Anthropocene is admittedly in part motivated by a sense of ecological grief that many environmentalists, like me, face as we ponder the level of climate change, pollution, coral mortality, and habitat and species loss we are experiencing. Surfers are also starting to see surf breaks as increasingly threatened by human-influenced ecological issues. It seems like almost daily (ok probably more like monthly) there is a news article in the surf media about how coastal development and infrastructure have either destroyed or are threatening to destroy a beloved surf break (Eckardt, 2024).

The limited research on how sea-level rise (SLR) might impact surf breaks also doesn't look good. Some suggest that, if you take California as a proxy, as many as 70% of our surf breaks might be threatened, if not extinct, in the next 50 years at current SLR projections (Reineman, Thomas, & Caldwell, 2017; Sadrpour & Reineman, 2023). There are surf-specific non-governmental organizations (NGOs) such as the Surfrider Foundation, Save The Waves Coalition, and Surfers Against Sewage fighting to raise awareness of the myriad threats facing surf ecosystems and working to mobilize support for preserving not only access but the very continual existence of the surf ecosystems many of us value.

Anything endangered or on the brink of extinction (surf breaks included) tends to acquire enhanced meaningfulness and value (Sodikoff, 2011). If I think a surf break will always be there, I will value it differently than if it is something that I think might vanish without action. Extinction events catalyze social change, yield new perspectives on temporality and existence, inspire creativity, propel technological advancement, and mobilize social movements (Sodikoff, 2011). The surf world is now mired with many examples of waves lost to port expansions and other hard structures and other waves on the knife-edge of that same fate (Eckardt, 2024). Endangerment, however, moves beyond objective facts; it is also an effectual experience with emotions (Vidal & Dias, 2015). Endangerment allows for the nostalgia and mourning of things that are not yet lost (Saroni, 2021).

In Bocas del Toro, Panama, where I live and work, I feel this daily. Research suggests that the world's most vulnerable waves are those breaking on coral reefs where hard structures, like sea walls, limit the ability of the coastline to retreat as sea-level rise (Sadrpour & Reineman, 2023). Paunch, the wave I surf every day there is surf, is very much endangered. Even without any talk of the potential impacts of the proposed infrastructure, local surfers have suggested that SLR has reduced the frequency of quality surf conditions (Mach, Rothrock, Stark, & Nahmias, 2024). They say larger swells are required for the waves to break and that once barely perceptible tidal changes can cause major differences in the conditions. In short, SLR is already decreasing the quality and consistency of the surf.

Currently, the road along the coast adjacent to where Paunch breaks is falling into the ocean. This allows wave energy to extend over the road when there are big swells. Because this limits refraction (the wave bouncing back to sea and messing with the surf), it is great for surfers, but not so convenient for anyone else trying to use a road to get anywhere past Paunch on the island. The proposed gabions, or hard structures to stabilize the coast and allow for cement coastal road construction, could threaten this tenuous balance. Isla Colon, the island where this wave is situated, has already experienced the beginning of a road project meant to stabilize the road in the face of SLR. Waves crest the cement barriers and refraction happens everywhere during big swell events in this new section of road. This new infrastructure stops right at the point where our surf breaks lie, but there are plans to construct the road all the way out to our northern beaches. This plan could have grave consequences on all our island's surf breaks, which we have estimated to bring in roughly 37 million USD per year in surf tourism-related expenditures (Segnitz, Kishimoto, Lynch, & Mach, 2025).

The issue is not easy; there are already hotels, houses, and restaurants along this road that could be cut off from town and the airport if nothing is done to stabilize the road. Many people's cars have also been destroyed (or are in the process of decay like my own) by the rutted-out road and river crossings needed to traverse it. But stabilizing the road could also prohibit the ability to surf, which is the lifeblood of the economy in that region and the inspiration behind a lot of the in-migration and development (Mach, et al., 2024).

The construction' firm say that they have technical strategies to ensure refraction does not impact the surf. But will it work? Is it worth the risk? Can we stop it? What if we don't, and the waves are lost? Could they be brought back? Would anyone bother to bring them back? These are all questions that linger in our minds here in

Bocas, but this sentiment is shared by surfers around the world who are united by this vulnerability.

Surf breaks, I sense, invite a sort of anticipatory nostalgia (Choy, 2011) or a kind of mourning for the loss of certain types of experiences in certain types of places that are not yet gone. Many of us have experienced older surfers mourning the loss of the glory days of surfing, way back before too many of us were doing it. For them, the waves haven't changed necessarily, but the experience of surfing has been fouled by too many newcomers who do not understand the ethics and codes forged and negotiated over time (Mixon, 2014). Others mourn the days when you could go on a surf trip and find uncrowded waves in towns that have not yet morphed into homogenous and efficient surf travel markets with any type of lodging and services any surfer of any level might require during a brief visit. Some say surf cameras are ruining their spots by bringing too much attention and causing crowding whenever internet surfers see quality surf opportunities on their screens (Mann, Warren, & Baird, 2023).

In short, even without potential threats to the viability of surf ecosystems to provide waves, surfing, like many other subcultures, has many who are crotchety and mourn aspects of the real or imagined past. In that sense, as we ponder the future of surfing and the role of technology in it, we want to be mindful of what types of experiences we are hoping to sustain and for whom.

There are tremendous benefits to be gathered from surfing, and the more people doing it can help create a society where people care more about coastal spaces (Mach & Ponting, 2022). When a proposed pier project threatened the viability of my home break at Paunch, the community mobilized to stop the bulldozers by holding up signs in defiance—locals became reef huggers, literally standing between the construction crew and the corals. With amplifications from Kelly Slater (the GOAT) and Dr. Chad Nelson (CEO of Surfrider Foundation), the interwebs enhanced the watchdog function, and the issue was exposed to the global surf community. Thousands of surfers eventually signed an online petition in support of stopping the project in solidarity with the local surfers. I am confident that this spotlight was important in halting the project. This issue, however, and many others like it, is complicated.

In this case, for example, the developer said the pier project was meant to protect coral from being trampled by organizing and concentrating where surfers enter and exit the water. Watching people walk all over the reef every day makes me somewhat sympathetic to this claim. They also suggested the reef they proposed to build over was already dead—a claim that was never adequately confirmed nor denied by the scientific community on the archipelago. The important thing to point out is, however, how the global connectivity of surf enthusiasts can act as a conduit for forcing deliberations on matters of importance and slow technological progress in the furtherance of having dialogue surrounding key issues.

This brings us back to the main goal of this book: to spark a general dialogue about surfing in the Anthropocene and get people imagining.

The Coevolution of Technology, Environment, and Society

Now to the organization of this volume. Rather than seeing technology as deterministic, I think it is best to view technology, environment, and society (TES) as coevolving (Norgaard, 1994; Mach, 2017). Any one of these analytical categories can exert pressure on others, but at the end of the day, there are feedback loops, and they morph in response to one another to comprise the contemporary milieu.

The example that always comes back to my mind to illustrate this coevolution in the context of surfing is the comparison of two of the world's most wave-rich islands—Bali, Indonesia, and Oahu, Hawaii. I would not venture to debate which island has better surf. Both have multiple world-class waves and boast their own "miracle miles." The more important consideration is what inspired Hawaiians to invent the first surfboard (technology), begin surfing in the 5th century, and then build the practice so firmly into their societ'y structure, while no one surfed in Bali until the 20th century.

This illustrates that the environment (one amenable to high-quality surfing) does not on its own determine whether the surrounding society will invent something like surfing or take it up in some way, shape, or form. In Bali, long-standing cultural beliefs instilled a deep-seated fear of the ocean. The sea was conceptualized as the farthest location from the mountain tops where the gods were situated, and the ocean was seen as the dwelling place of unknown dangers (Moore, 2010). The Balinese traditionally relied more on agriculture and very near-shore fisheries and did not develop any ocean-based leisure activities, even close to surfing.

Those on Oahu, however, also occupied warm wave-filled waters in a geographically isolated region (environment). They, however, had established protein surpluses through advanced aquaculture and agriculture and vessels for navigating in the ocean (Technology) and believed in close commune and interaction with their gods embodied in nature and valued leisure within their Kapu system that organized behavior (Society) (Westwick & Neushul, 2013; Mach, 2014). The point is that TES relations are crucial to defining which technologies have been invented (or adopted) and used in any time/space in question. This was true as far back as surfing's origin and remains the case today despite how much more complex things may appear.

Chapter contributions for this volume will be categorized under the three parts of this book: (1) Technology, (2) Environment, and (3) Society, based upon the central focus of each contribution. This is done for analytical clarity and organizational purposes. Each chapter examines the TES interconnections in some way, so I do not wish for readers to miss this solely based on where they were placed in the volume. Each author is, furthermore, an expert in the topic area they address, and this is why I am thrilled to be able to share their ideas on the past, present, and future of surfing in the Anthropocene.

Technology

The book will begin with the Technology part, and the main themes that will be covered are surf parks, wetsuits, and surf cameras. The opening chapter was written by Jess Ponting, the director of the Center for Surf Research and a co-founder of Surf Park Central. His chapter discusses the history of surf pool technology, the pace of acceptance and adoption of commercial surf parks around the world, and the new social and recreational spaces that are emerging as a result. Jess envisions a world where surf parks do not replace ocean surfing but allow wider access to the benefits of wave riding, controlled and safe conditions for learning, opportunities to progress much faster in predictable settings, and new planned destinations combining surf parks with lodging and other touristic services. He also refutes the idea that surf parks are inherently wasteful in their water and energy use and argues that surf parks can be sustainably developed and provide net positive environmental benefits with conscious planning.

On the other hand, a group of scholars from the University in Angers, France, led by Prof. Christophe Guibert, discuss the political and ecological debates surrounding surf pools in France. Christophe, Carlo, Jules, and Yann lead us in a thoughtful discussion as they consider why there are currently no surf pools in France's Surficon Valley (i.e., the Casets area where France's best surf breaks and most surf industry organization's offices are located) despite numerous proposals and high-profile boosters for these installations. Without ignoring the proposed benefits, they show how diverse coalitions surrounding proposed installation sites have energetically opposed the projects, citing issues related to waste (water and energy), traffic, congestion, and noise. Many have also argued for the philosophical desire to keep surfing in the ocean in France and seemed to take pride in the prospect of keeping the country free of surf pools despite their widescale adoption in nearby countries.

Combined, these chapters give us a lot to think about in terms of how we might think about surf pools and parks and the roles they might play in the future of surfing in the Anthropocene. Given the high level of investment in this space, the use of AI in design, and the demand for surfing experiences, it is likely that these experiences will proliferate. But where will the horizon be for surf parks? Will certain geographies become defined by their opposition? Will enough parks be profitable to reach scale? Will growth be exponential, or will these investments fizzle out as fad after failing to find a sustainable market? Only time will tell.

In Chap. 3, Sean C. Newcomer and Jeff A. Nessler, surf scientists from Cal State San Marcos' Kinesiology Department, along with Tyler Wiles from the University of Nebraska at Omaha, discuss every surfer's favorite love-to-hate item, wetsuits. This chapter will explore the history, current technological trajectories, and potential implications of efforts to improve the efficiency, environmental friendliness, functionality, and effectiveness of the suits that keep us warm while facilitating longer surfing sessions in more places worldwide. They showcase how, from their inception, until recently, science has rarely been utilized to empirically improve the performance of wetsuits. As a result, thickness has been added to the wrong places and materials remain toxic for most brands. But the message is hopeful. With the promise of more scientific integration in R&D, they help us visualize a future where wetsuits are comfortable, light, and warm while also being healthy for our bodies and surf ecosystems. They might even be designed to increase paddling speed and support areas where surfers tend to have chronic injuries like the knees (in my case, please hurry on this later part). Their quest into the future is interesting and helps us all to envision potentially drastic changes in what we use to keep us warm—including potentially biodegradable layers of slime.

Chapter 4, written by Tyler Poisson, a doctoral candidate in Communications at UMass Amherst, takes on the ubiquity of surf cameras. He does not ask whether surfers should blindly accept cameras or smash them to pieces, but instead, he asks, what access to them does to our sense of and experience of time in the Anthropocene. It is quite a fascinating read and gives us all new things to consider as we navigate surf camera access and how they might shape what we actually do in the world and how we experience it as nature becomes ever more mediated by technology.

Environment

Part II will focus on environmental themes such as the impacts of climate change on surfing in unconventional non-ocean environments, wave saving technology, shark mitigation technologies, plastic and wastewater pollution. Scott Laderman, Professor of History at the University of Minnesota Duluth and author of Empire in Waves, opens this part with what I believe might be the first writing on how climate change and other environmental factors are impacting fresh and brackish water surfing in lakes and rivers. His chapter discusses the growing popularity of unconventional surfing, as inland surfers seek the same joy and benefits from the activity as their coastal counterparts. He cautions us, however, not to take these experiences for granted and details various vulnerabilities these often-neglected surfing ecosystems are facing in the Anthropocene.

Chapter 6 is by Diego A. Sancho-Gallegos, the Stewardship and Resilience Manager at Save The Waves Coalition. His writing helps us to understand the opportunities and challenges associated with utilizing technology (e.g., Apps and AI) in the fight to conserve surf ecosystems in the face of myriad challenges. As someone who has been on the front lines of arming conservation organizations with useful and valuable information, his commentary on this topic will be interesting to anyone who has any interest in saving waves.

Daryl McPhee is an Associate Professor at Bond University in Australia and a fisheries expert. His chapter walks us through the history of mitigating shark attacks on surfers. He then goes on to present empirical data on past unprovoked shark attacks on surfers before discussing the current findings related to the efficacy of different technological efforts to limit unwanted encounters. He concludes with some interesting future proposals for mitigating risk and reducing harm in sharky seas.

The next chapter is written by the Surfrider Foundation's California Policy Associate Director. Therein, Jennifer Savage writes about the ubiquity of plastics in and around our surf breaks and the environmental and psychological harm this causes. She writes about her direct experience in the ongoing legal and political fights to reduce single use plastic creation at the source and explains why this is the most essential pathway for ocean health. The urgency is clear, and the hope is that we can select for an ocean where peak plastic is in the rear view and we can begin reducing its volume in the ocean.

The final chapter in this part, by David Brown, is also about ocean pollution. He is stormwater specialist for the City of San Diego and writes about his polluted leisure and occupational experiences (both as a surfer and lifeguard on San Diego's beaches) before diving into his experience researching and evaluating the health risks of run off for city government. I'll admit, as someone who spent a few years living in San Diego, the problem is grosser and larger than I ever thought. David's experiences, research and ideas related to this important topic are critical to understanding the scope of the problem, methods for mitigating the impacts of coastal pollution on our health, and potential solutions for better monitoring and limiting pollution in the ocean.

Society

The final part will look at societal issues such as the future of psychedelics in surf culture, the opportunities and challenges involved with using surfers in citizen science projects, technology and coastal access and the future of surf magazines. I first saw Jarrett Rose, an Assistant Professor at SUNY, speak about the history of psychedelics in surf culture at the Surf and Skate Collaborative Conference at San Diego State University in 2023. As soon as I knew I was putting this book together, I reached out to him for his thoughts on what he thinks about psychedelics in contemporary surf culture and how things might evolve into the future. This chapter is his answer. It provides a very interesting commentary on the transition from counterculture to mainstream psychedelic use in surf culture for healing and self-improvement.

Chapter 11 is dedicated to Data Bitches', citizen science, and sea-level rise. In this chapter, Brie Iatarola from the University of California San Diego recounts her firsthand experience participating in a citizen science project at her dissertation field site, Trestles. This is the future site that will host the Olympic games in 2028 and has hosted many other World Surf League events and finals over the years. That is to say, the impacts of SLR at this beach can be considered outsized in the surf world. In this chapter we learn about the challenges involved with leveraging the willingness of concerned citizens to support scientific endeavors in beloved surf zones. Brie leaves us with recommendations for boosting training for data bitches, enhancing lines of communication and limiting data loss, which can help improve this form of scientific collaboration in an era of potentially reduced research funding for vital environmental monitoring projects.

In Chap. 12, Jessie Reiblich, an attorney and Assistant Professor at the University of Rhode Island, walks us through the intricacies of coastal access for surfers. In sort, we cannot surf what we cannot access. In this chapter, we learn about how sea-level rise, localism, and the privatization of coastal land can impact surfer's ability to pursue waves in different contexts around the United States. He envisions how technology can be utilized to monitor changes in coastal access and help arm surfers with the information they need to ensure surf ecosystems remain open to current and future generations of surfers.

The last chapter was written by Craig Sims, an Assistant professor at Bond University. He writes about the history of surf magazines as a defining and enduring technological artifacts of surf culture. While many prophesized the death of the print magazine amidst the furry of surf related social media content providers and influencers, Craig writes about why after only a brief dip, the surf magazine is having a renaissance. His data shows that surfers trust print content more and like that the content is not as ephemeral as what is posted digitally. These are major reasons why he argues magazines are likely to remain critical to articulating the many facets of our evolving surf culture as we float forward in time.

I truly hope you enjoy the content that follows. I believe it will allow you to peer into the future of surfing in the Anthropocene and help you to develop new questions and strategies for navigating what is to come!

Leon Mach

References

Angé, O., & Berliner, D. (2014). *Anthropology and nostalgia*. Berghahn Books, Inc.

Barilotti, S. (2002). Lost Horizons: Surfer colonialism in the 21st Century. *The Surfer's Journal, 3*(11).

Bell, D. (1976). *The cultural contradictions of capitalism*. Basic Books.

Britton, E., Kindermann, G., Domegan, C., & Carlin, C. (2020). Blue care: A systematic review of blue space interventions for health and wellbeing. *Health Promotion International, 35*(1), 50–69. 10.1093/heapro/day103

Choy, T. (2011). *Ecologies of comparison: An ethnography of endangerment in Hong Kong*. Duke University Press.

Eckardt, J. (2024, February 6). Watch: Epic waves facing extinction and what surfers can do to help. *Surfer Magazine*. Retrieved March 26, 2025, from https://www.surfer.com/news/epic-waves-facing-extinction-surfers-help

Fodor's (2024, November 13). Fodor's No List 2025. *Fodor's*. Retrieved March 28, 2025, from https://www.fodors.com/news/news/fodors-no-list-2025

Ghosh, R. (2024). A fond farewell to the anthropocene. *Issues in Science and Technology, 40*(3), 20–22. 10.58875/OUAY7538

Gibbs, K., Wilkie, L., Jarman, J., Barker-Smith, A., Kemp, H., & Fisher, Z. (2022). Riding the wave into wellbeing: A qualitative evaluation of surf therapy for individuals living with acquired brain injury. *PLOS One, 17*(4). 10.1371/journal.pone.0266388

Hill, L., & Abbott, J. A. (2009). Representation, identity, and environmental action among florida surfers. *Southeastern Geographer, 49*(2), 157–170.

Irvine, R. (2021). Peaceful countryside: Ecologies of longing and the temporality of flux in contemporary Mongolia. In O. Angé & D. Berliner (Eds.), *Ecological nostalgias: Memory, affect and creativity in times of ecological upheavals* (pp. 145–165). New York, NY: Berghahn Books.

Kushner, D. (2023, May 2). The Surf Bros, the Villagers, the Wave Doctor, the Tech Money, and the Fight for Fiji's Soul. *Vanity Fair*. Retrieved from https://www.vanityfair.com/style/2023/05/fiji-surf-wars

Laderman, S. (2014). *Empire in waves: A political history of surfing*. University of California Press.

Laderman, S. (2017). A world apart: Pleasure, rebellion, and the politics of surf tourism. In D. Z. Hough-Snee & A. S. Eastman (Eds.), *The critical surf studies reader* (pp. 47–61). Durham: Duke University Press.

Lawler, K. (2011). *The American surfer: Radical culture and capitalism.* Routledge.

Mach, L. (2014). *From the endless summer to the surf spring: Technology and governance in developing world surf tourism.* University of Delaware. Proquest.

Mach, L. (2017). Surfing in the technological era. In D. Borne & J. Ponting (Eds.), *Surfing and Sustainability* (pp. 41–71). New York: Routledge.

Mach, L., Rothrock, E., Stark, S., & Nahmias, J. (2024). Place attachment, wellbeing, and conservation in surf destination communities. *Frontiers in Sustainable Tourism, 3.* 10.3389/frsut.2024.1387081

Manero, A., & Mach, L. (2023). Valuing surfing ecosystems: an environmental economics and natural resources management perspective. *Tourism Geographies.* 10.1080/14616688.2023.2261909

Mann, M., Warren, I., & Baird, A. (2023). Surfveillance: SurfCams, social issues and regulatory intervention. *Journal of Sport and Social Issues, 48*(3–4). 10.1177/01937235241276821

Mixon, F. (2014). Bad vibrations new evidence on commons quality and localism at california's surf breaks. *International Review of Economics, 61*(4), 379–397.

Noble, D. (1983). Technology's politics: Present tense technology. *Democracy, 3*(2), 8–24.

Norgaard, R. B. (1994). *Development betrayed: The end of progress and a coevolutionary revisioning of the future*. Routledge.

Ponting, J., Teixeira, A., Costa, J., Vasconcelos, J., & Mach, L. (2023). Refining the framework analysis for sustainable surf tourism (FASST) for use in High-Income country settings: The case of Peniche, Portugal. In M. Roberts & K. Lawler (Eds.), *Roll and flow: The cultural politics of skateboarding and surfing* (pp. 283–303). San Diego State University Press.

Reineman, D., Thomas, L., & Caldwell, M. (2017). Using local knowledge to project sea level rise impacts on wave resources in California. *Ocean & Coastal Management, 138*, 181–191.

Sadrpour, N., & Reineman, D. (2023). The impacts of climate change on surfing resources. *Shore and Beach, 91*, 32–41.

Saroni, A. (2021). *Starfish encounters: Endangerment ethnography in Boca del Drago, Panama.* San Antonio: Univeristy of Texas.

Sodikoff, G. M. (2011). *The anthropology of extinction: Essays on culture and species death.* Indiana University Press.

Stranger, M. (2011). *Surfing life: Surface, substructure and the commodification of the sublime.* Ashgate.

Vidal, F., & Dias, N. (2015). *Endangerment, biodiversity and culture.* Routledge.

Wapner, P. (2020). *Living through the end of nature: The future of American environmentalism.* MIT Press. 10.7551/mitpress/8454.001.0001

Westwick, P., & Neushul, P. (2013). *The World in the Curl: An unconventional history of surfing.* Crown Publishing.

Wheaton, B., Waiti, J., Olive, R., & Kearns, R. (2021). Coastal communities, leisure and wellbeing: Advancing a trans-disciplinary agenda for understanding ocean-human relationships in Aotearoa New Zealand. *Public Health, 18*(2), 450. 10.3390/ijerph18020450

Contents

Part III Society

Contributors

David L. Brown Independent Researcher, San Diego, CA, USA

Carlo Grigoletto University of Angers, Angers, France

Christophe Guibert University of Angers, ESO Laboratory (UMR CNRS 6590), Angers, France

Briana M. Iatarola University of California San Diego, San Diego, USA; San Diego State University, San Diego, USA

Scott Laderman University of Minnesota, Duluth, USA

Daryl McPhee Faculty of Society and Design, Bond University, Robina, QLD, Australia

Jeff A. Nessler Department of Kinesiology, California State University San Marcos, San Marcos, USA

Sean C. Newcomer Department of Kinesiology, California State University San Marcos, San Marcos, USA

Tyler Poisson University of Massachusetts Amherst, Amherst, MA, USA

Jess Ponting Recreation and Tourism Management L. Robert Payne School of Hospitality & Tourism Management College of Professional Studies and Fine Arts, San Diego, CA, USA

Jules Raimondo University of Angers, ESO Laboratory (UMR CNRS 6590), Angers, France

Jesse Reiblich Assistant Professor, Department of Marine Affairs, University of Rhode Island, Kingston, USA

Jarrett Robert Rose Community & Behavioral Health Program and the Department of Sociology, State University of New York (SUNY) Polytechnic Institute, Utica, NY, USA

Diego A. Sancho-Gallegos Save The Waves Coalition, Santa Cruz, CA, USA

Jennifer Savage Surfrider Foundation, Manila, CA, USA

Craig P. Sims Bond University, Gold Coast, QLD, Australia

Yann Vilboux University of Angers, Angers, France

Tyler M. Wiles Department of Biomechanics, University of Nebraska at Omaha, Omaha, USA

Part I
Technology

Chapter 1
The Future of Surf Parks

Jess Ponting

Five years. 10 million waves. Over 300,000 guests have surfed the distance from the Earth to the Moon. 102,450 people have learned to surf at Australia's first surf park.
URBNSURF CEO Jennifer Vandekreeke, LinkedIn Post: April 15th 2025.

Introduction

Projecting the future trajectory of the surf park industry involves a high degree of uncertainty, and this chapter does not claim to offer definitive predictions. What follows is a mix of empirical data, early industry evidence, and, where necessary, best-guess projections grounded in what we know today. It's also written with a clear bias: toward sustainability. That bias isn't just philosophical (though it is that too); it's practical. The long-term success of the surf park industry depends on how well it aligns with community values, environmental stewardship, and cultural authenticity—whether operators realize it or not.

In the pages ahead, we'll trace the origins of surf parks from novelty attractions to a fast-growing global industry. We'll look at the technologies powering modern parks, the shifts in business models that are attracting serious capital, and the cultural, environmental, and social implications of a surfing experience untethered from the ocean. Academic studies and user surveys offer clues about how the public perceives these facilities and what expectations they carry—particularly around sustainability. We'll also explore how surf parks might be harnessed to protect the very ecosystems that inspired them. And finally, we'll make some educated guesses about where this industry is headed by 2040.

J. Ponting (✉)
Recreation and Tourism Management L. Robert Payne School of Hospitality & Tourism Management College of Professional Studies and Fine Arts, San Diego, CA, USA
e-mail: jponting@sdsu.edu

L. Mach (ed.), *The Future of Surfing in the Anthropocene*,
https://doi.org/10.1007/978-3-032-09377-6_1

How Did We Get Here?

In 2026, there will be at least twenty-five operational dynamic wave surf parks[1] and more than thirty deepwater standing wave attractions.[2] This chapter will focus on the dynamic wave sector of the commercial surf park industry, the origins of which can be traced back to a hydraulic flap system capable of producing gently spilling waves in swimming pools that debuted at the 1912 International Hygiene Exposition in Dresden, promoted as a therapeutic, 'enhanced swimming experience' (Westwick & Neushul, 2013). Soon after, hydraulic flap enhanced swimming pools sprung up across Europe and eventually crossed the Atlantic to the United States. Somewhere in the realm of 500 were built by the end of the century (Warshaw, 2004). Surfing in a wave enhanced swimming pool began in Japan in 1966, when a venue called Surf-a-Torium allocated 15 min each hour for surfers to ride gentle, tiny, spilling waves. The first dedicated commercial surf lagoon, 'Big Surf' opened in 1969 in Tempe, Arizona. The 300 × 400-foot lagoon utilized a 40-foot water drop system to generate waist-high, spilling waves (Westwick & Neushul, 2013).

In the early 1970s, the Offshore Technology Corporation (OTC) which built wave testing models for oil platforms recognized recreational applications of their technology and spun off Surf Tech, a new venture focused on creating surf pools. Their first project, the Aquaboggan water park in Saco, Maine, produced modest results, but their next effort at Dorney Park Wildwater Kingdom in Pennsylvania garnered more attention. In 1985, Dorney Park hosted the World Professional Inland Surfing Championships. Unfortunately, the lackluster wave quality meant the event failed to captivate the viewing public or the surfers themselves (Lowry, 1985; Warshaw, 2004).

The quality of surf pools took a major leap forward with the opening of Typhoon Lagoon at Disney World Orlando in 1989, a surf lagoon capable of producing a small but perfectly shaped plunging wave and available to rent outside of normal Disney operating hours. In 1993, Ocean Dome opened in Miyazaki, Japan, as part of the Sheraton Seagaia Resort. This $300 million development included an advanced wave-generating system built by Mitsubishi Heavy Industries, better known for building nuclear reactors, that produced waves that would take another 20 years to be topped in terms of quality. Ocean Dome closed in 2007 due to an inability to turn a profit against the massive capital investment required to bring it to life.

The nascent surf park industry was dealt a significant blow in 2008 with the failure of the Ron Jon Surf Park. The project brought together many of the best minds in surf park development including the company Artificial Surfing Reefs that had mapped the bathymetry of many of the world's best surf breaks and hoped to be able to re-create

[1] A dynamic wave, or travelling wave, in the context of surf park refers to a wave that breaks across a large lagoon with the surfer moving in some cases hundreds of meters.

[2] Deep water standing waves are based on the hydraulic jump phenomenon that occur in rivers and were first described by Leonardo DaVinci. A pneumatic jump in the Eisbach (Ice River) in central Munich Germany inspired the development of artificial replicas of this experience. The wave is stationary but allows surfers to ride conventional surfboards and surf back and forth along the wave.

them with a programmable robotic pool floor. A highly successful pre-development phase proved there was demand for the Ron Jon Surf Park, established financial viability for the proposed model, and raised sufficient capital ($10 M) to execute the project. Unfortunately, the pool itself was a spectacular failure—the first wave ripped the robotic floor off the bottom of the lagoon and destroyed it. The project was abandoned. This catastrophic failure commingled with the global recession of 2008 and led to a pause in surf park financing that would last for 7 years. The only exception was the Wadi Adventure surf lagoon in the United Arab Emirates which opened in 2012.

Wavegarden Lagoon (which is now functionally obsolete) and Kelly Slater Wave Co are based on different applications of the same basic concept: mechanical foils being dragged through the water. Relatively few wave variations can be attained beyond dragging the foils faster or slower. Alternatively, Surfloch, American Wave Machines, and Endless Surf all use sequentially fired pneumatic caissons to generate waves and come with the distinct advantage of almost infinite wave variation and the capability of switching between wave types from one wave to the next. Wavegarden Cove is an electromechanical system that uses electric motors to move paddles that generate waves on either side of a central pier every 4 s, a very efficient system which also allows for significant wave variation. Swell MFG also uses an electromechanical system to displace water. All Waves uses a giant inflatable pillow under the water to displace water from below while Surf Lakes drops a large weight into a lake to produce outward moving circular waves from a central point that are shaped and broken in a variety of ways using variations of bathymetry. The Surf Lakes system is capable of some wave variation provided by static bathymetry and varying the size and number of waves being produced.

Several other novel wave generation technologies are also in the testing and prototyping phases of development. For example, French-based 'Okahina Wave' has developed a model that requires no concrete, no walls, and no basin floor bathymetry in its construction. Instead, a circular module 100 m in diameter is partially submerged in an existing body of water. The module contains a proprietary wave-making system causing swells to break around the circumference, taking 45 s to circumnavigate the central module. Three waves break simultaneously equidistant apart from each other. While this model does require the construction of a bridge from land out to the center of the module in the water, it is demountable meaning the entire surf park can be potentially installed for an event, or a season, and then removed with no permanent structures required. While a 1/10 scale model is the largest iteration built to date, the concept has apparently won the support of some of France's conservation community who have become notorious in their opposition to surf park development. See a rendering of the technology in Fig. 1.1.

The first commercial surf park utilizing technologies more advanced than simple flap and water drop systems to market was Surf Snowdonia. This was a Wavegarden Lagoon project in Conwy, North Wales that opened to the public on August 1st 2015. The facility attracted 15,000 visitors in its first month of operation and staged a successful major international competition in its second month. A breakdown in September 2015 caused a six-day closure before a major mechanical failure and tear

Fig. 1.1 Rendering of an Okahina wave installation. *Source* okahinawave.com

in the lake liner forced the park to close for the winter 8 weeks early. Though it did produce a shoulder high spilling wave every 90 s, the six foot plunging wave every 60 s promised in pre-opening media releases was never achieved (Temperley, 2015). Surf Snowdonia reopened on March 19th 2016 with a marginally improved product as a result of work done on its northern distributive shores to hasten the dampening of wave energy. A second, larger, Wavegarden Lagoon project, NLand, was completed in 2016 in Austin, Texas and finally opened after considerable legal wrangling with its county government over water filtration requirements. NLand closed for good in November 2018 and was purchased by the World Surf League in January of 2019 and is now the site of a 190 acre amenitized residential real estate development with more than 140 condominiums, a KSWaveCo facility *and* an Endless Surf lagoon. Surf Snowdonia, which had been renamed Adventure Parc Snowdonia, announced it was closing its surfing operations in September of 2023 due to ongoing mechanical issues and an inability to reach a viable preventative maintenance arrangement with its technology provider Wavegarden.

Wavegarden Cove, which was unveiled in prototype form in 2016 is theoretically capable of producing almost 1000 very high-quality waves per hour in a relatively small footprint. Wavegarden has a world-leading ten operating commercial Cove facilities outside of its test facility in Basque Country, Spain. The Wave Bristol was first to market opening in 2019, URBN Surf in Melbourne Australia opened in January 2020 (then closed and opened several more times due to Covid-19 pandemic lockdowns), and Wave Park in Siheung, South Korea opened in October 2020. Wave Park, in addition to its waves and facilities being powered and its water consistently heated to 70 °F/21 °C (even when it is snowing) by renewable energy generated at a local biowaste facility, has also innovated for the Korean market by introducing an almost minute long slow-moving sloping wave suitable for South Korea's emerging longboard scene. See Table 1.1 for a complete listing of all dynamic wave surf parks and their wave generation technologies at the time of writing.

Just four months after the opening of Surf Snowdonia, in December 2015, KSWaveCo revealed its first waves at the Surf Ranch in Lemoore in California's Central Valley, south of Fresno. Operating more than three years before the first full sized commercial Wavegarden Cove opened, the KSWaveCo Surf Ranch—now the WSL Surf Ranch since the WSL acquired KSWaveCo in mid 2016—was a quantum shift in the way the world viewed surf parks. Previously surf park waves were compared with waves in nature, the best to that point had been Surf Snowdonia which was being compared favorably with a small day at Greenmount on Australia's Gold Coast. The reveal of the KSWaveCo facility turned this on its head and now the world's most perfect waves are compared favorably with surf parks. While it produces the largest (2 m), longest (almost 1 km, ride time approaches 1 min), and most perfect waves of any technology, these facilities require enormous physical footprints—more than 15 acres—before any ancillary buildings and only produce one wave every four minutes.

American Wave Machines (AWM) has four operational surf parks. BSR Surf Ranch in Waco Texas opened in June 2018 and immediately made an impact with the variability of its waves and clear benefits for high performance surfing and training, particularly technical aerial maneuvers. Unfortunately, within a year there had been two fatalities at the BSR Surf Resort, it is important to note that both were unrelated to the AWM wave technology. One death was caused by a brain-eating amoeba, *Naegleria fowleri*, that occurs in natural stagnant warm water environments and insufficiently treated/filtered water in human-made water attractions. The second fatality involved the drowning of a swimmer in the surf lagoon during a night surfing session. Both incidents highlight the importance of water quality and operational safety protocols. A second smaller AWM facility (1.5 acres) opened inside the American Dream mall in New Jersey, pushing its first waves in November 2019. Facilities have subsequently opened in Japan and Brazil (see Fig. 1.1).

Surf Lakes pushed its first waves in October of 2018 at its full scale research and development facility in Yeppoon in North Queensland, Australia. The facility has been gradually ramping up the size of the waves it is producing and in October of 2020 at Surf Park Summit released video of extremely high-quality surf. Surf Lakes can, in theory, produce more than 2000 wave per hour and a maximum wave height of 2.4 m. In 2021 Surf Lakes claimed to have 49 'legitimate projects' in its pipeline across 9 countries, though at the time of writing none of these have reached the stage

Table 1.1 Surf park industry milestones 1912–2025

Year	Milestone	Type	Location	Technology	Status
1912	Flap system showcased	Prototype	Dresden	Flap system	N/A
1966	Surf-A-Torium	Surf pool	Japan	Flap system	N/A
1969	Big Surf	Surf lagoon	AZ, USA	Water drop	Closed, 2020
1985	Inland Surfing Championships	Contest	PA, USA	Flap system	N/A
1989	Typhoon Lagoon	Surf lagoon	FL, USA	Barr and Wray	Active
1993	Sea Gaia Ocean Dome	Surf lagoon	Japan	Mitsubishi	Closed 2007
2008	Ron Jon Surf Park	Prototype	FL, USA	Custom	Failed
2012	Wadi Adventure	Surf lagoon	UAE	Murphy's waves	Active
2013	Wavegarden Lagoon	Prototype	Spain	Wavegarden lagoon	Obsolete
2013	Surf Park Summit	Conference	CA, USA	N/A	N/A
2015	Surf Snowdonia	Surf lagoon	Wales, UK	Wavegarden lagoon	Closed 2023
2015	Surf Ranch	Surf lagoon	CA, USA	KSWaveCo	Active
2016	Surf Park Summit	Conference	FL, USA	N/A	N/A
2016	NLand Surf Park	Surf lagoon	TX, USA	Wavegarden lagoon	Closed 2018
2017	Surf Park Summit Europe	Conference	Wales, UK	N/A	N/A
2018	Waco Surf	Surf lagoon	TX, USA	AWM*	Active
2018	Surf Lakes	Surf lagoon	Australia	Surf lakes	Redeveloping
2019	The Wave, Bristol	Surf lagoon	UK	Wavegarden cove	Active
2019	Skudin Surf	Surf lagoon	NY, USA	AWM*	Active
2020	URBNSURF Melbourne	Surf lagoon	Australia	Wavegarden cove	Active
2020	Wave Park	Surf lagoon	South Korea	Wavegarden cove	Active
2020	Surf Park Summit	Conference	Online	N/A	N/A
2021	Alaia Bay	Surf lagoon	Switzerland	Wavegarden cove	Active
2021	Surf Stadium Japan	Surf lagoon	Japan	AWM*	Active
2021	Surf Park Summit	Conference	Online	N/A	N/A
2021	Praia da Grama	Surf lagoon	Brazil	Wavegarden cove	Active

(continued)

Table 1.1 (continued)

Year	Milestone	Type	Location	Technology	Status
2022	Surf Park Summit	Conference	CA, USA	N/A	N/A
2023	Surfland	Surf lagoon	Brazil	Wavegarden cove	Active
2023	Boa Vista Village	Surf lagoon	Brazil	AWM*	Active
2023	Surf Park Summit	Conference	CA, USA	N/A	N/A
2024	Dallas Developers Tour	Conference	TX, USA		
2024	RiF010	Surf Canal	Netherlands	Surfloch	Active
2024	Palm Springs Surf Club	Surf lagoon	CA, USA	Surfloch	Active
2024	URBNSURF Sydney	Surf lagoon	Australia	Wavegarden cove	Active
2024	O2 Surftown Muc	Surf lagoon	Germany	Endless surf	Active
2024	Surf Abu Dhabi	Surf lagoon	UAE	KSWaveCo	Active
2024	Lost Shores Surf Resort	Surf lagoon	Scotland, UK	Wavegarden cove	Active
2024	Surf Park Summit	Conference	CA, USA	N/A	N/A
2024	All Waves	Prototype	Belgium	All waves	Inactive
2025	Revel Surf	Surf lagoon	AZ, USA	Wave MFG	Active
2025	SRF Park TLV	Surf lagoon	Israel	Wavegarden cove	Active
2025	AZ Surf Park Developers Tour	Conference	AZ, USA	N/A	N/A
2025	Euro Surf Park Industry Forum	Conference	Germany	N/A	N/A
2025	Atlantic Park	Surf lagoon	VA, USA	Wavegarden cove	N/A
2025	Surf Park Summit	Conference	VA, USA	N/A	N/A

* American Wave Machines

of construction. The Yeppoon facility is in the process of being redeveloped in a commercial facility.

Several new wave generation technologies have emerged and reached commercial deployment in recent years, reflecting ongoing diversification in wave generation technology. Surfloch, a pneumatic wave technology provider, powers the Palm Springs Surf Club (PSSC)—a compact wave pool built on the site of the former Wet n Wild water park in Palm Springs—which pushed out its first waves in prototype format in March 2020 and reopened at full scale consistently following a basin expansion in 2024. Surfloch is also behind the dynamic wave installation at Rif 10, a surf park developed within a repurposed canal in Rotterdam that began operations in 2024. Endless Surf, a new surf park division of long-established water park firm Whitewater West, also utilizes pneumatic technology; its first operational facility launched in Munich in 2024. Meanwhile, Belgian company All Waves introduced a novel wave generation approach using a submerged pneumatic textile pillow system, successfully demonstrated at a full-scale test facility in Belgium in 2024. Finally, Swell Manufacturing's hydro-mechanical, electrically powered wave system is featured at Revel Surf, the central attraction of the Cannon Beach mixed-use real estate development in Mesa, Arizona. The surf park is positioned as a community catalyst and commercial anchor, with developers reporting that retail tenants at Cannon Beach are paying rents up to 50% above market rates (Weinstein, 2025). A second, similar Cannon Beach development is underway in McKinney, Texas. See Table 1.1 for a timeline of surf park industry milestones from 1912 to 2025.

The Quickening

As shown in Table 1.1 and Figs. 1.2 and 1.3, the dynamic wave surf park industry has transitioned from slow, uncertain beginnings to a period of rapid growth, especially since 2015 when the first of the new generation of surf parks came online and dispelled the fears of surfers around the wave quality on offer at surf parks. The number of openings and the cumulative growth curve illustrate a market that has moved past early skepticism and is now gaining serious momentum.

Several factors have shaped this trajectory. Early surf park projects faced considerable challenges due to long development timelines (typically five years or more), the immaturity of wave-generation technology, and significant difficulty accessing capital. These pioneering projects, like Surf Snowdonia in Wales and NLand Surf Park in Texas, were made possible primarily by high-net-worth individuals investing personal or family money. Keen surfers within the Ainscough family (of Ainscough Crane Hire) and the Coors family (of Coors Brewing) leveraged family wealth to fund these ventures at a time when institutional financing was unavailable. These early parks proved that the concept could attract visitors. Surf Snowdonia, for instance, welcomed over 70,000 guests annually. However, both suffered from operational challenges, equipment failures, high maintenance costs, and ultimately closed. Wavegarden Lagoon technology became obsolete with the release of Wavegarden Cove technology. The Lagoon technology was prone to frequent breakdowns and had a throughput a fraction of the throughput of Cove technology and produced waves of inferior quality. With more robust technology on offer providing high-quality wave counts orders of magnitude higher than the previous technology, technology driven closures are not anticipated moving forward.

As the industry has matured, so too have its business models. Rather than relying solely on ticketed surfing with modest onsite FnB as a revenue streams, new projects

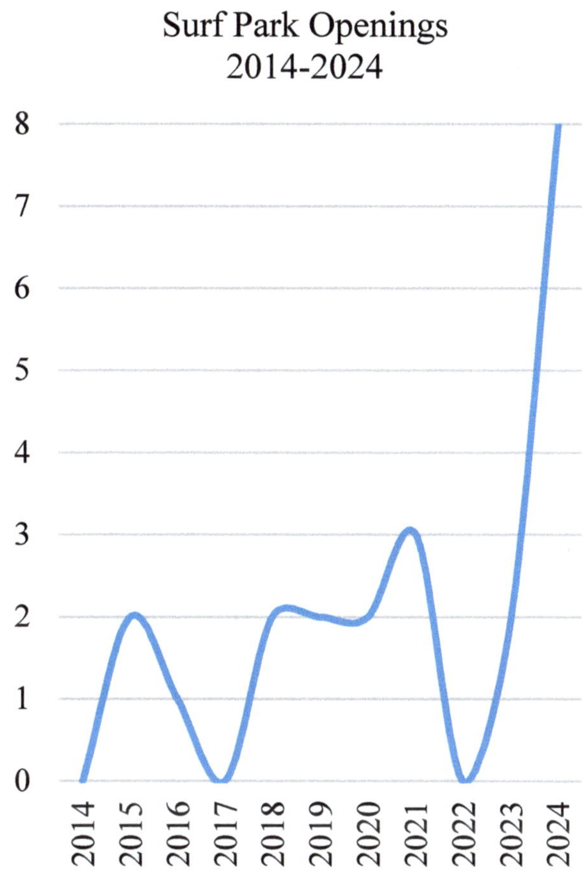

Fig. 1.2 Surf park openings, actual and cumulative 2014–2024. *Source* Ponting (2025) Presentation to Arizona Surf Park Developers Tour

are increasingly embedded as anchors for mixed-use real estate developments. Some developments leverage surf lagoons as public amenities driving food and beverage, retail, and hospitality offerings, while others anchor private residential communities, similar to residential golf course developments. In Brazil, developers have even sold residential lots pre-construction, using the proceeds to fund lagoon construction. The announcement of a surf park development on one greenfield site in Brazil resulted in an almost immediate tenfold increase in land values, underscoring the substantial economic uplift these projects can generate, according to Sean Young, Head of Development Projects at Wavegarden, speaking at the 2023 Surf Park Summit.

This evolution in strategy is supported by improved economic modeling, a growing body of operational data, and the emergence of surf-park-specific professionals—from feasibility consultants to booking software providers and AI video analysts. Developers like Beach Street Development and Operations, KSM Real Estate, Aventuur, URBNSURF, The Wave, Cannon Beach, and others including the teams behind Palm Springs Surf Club and Lost Shore Surf Resort are leading multi-park rollouts, both domestically and internationally. Institutional investors are beginning to take

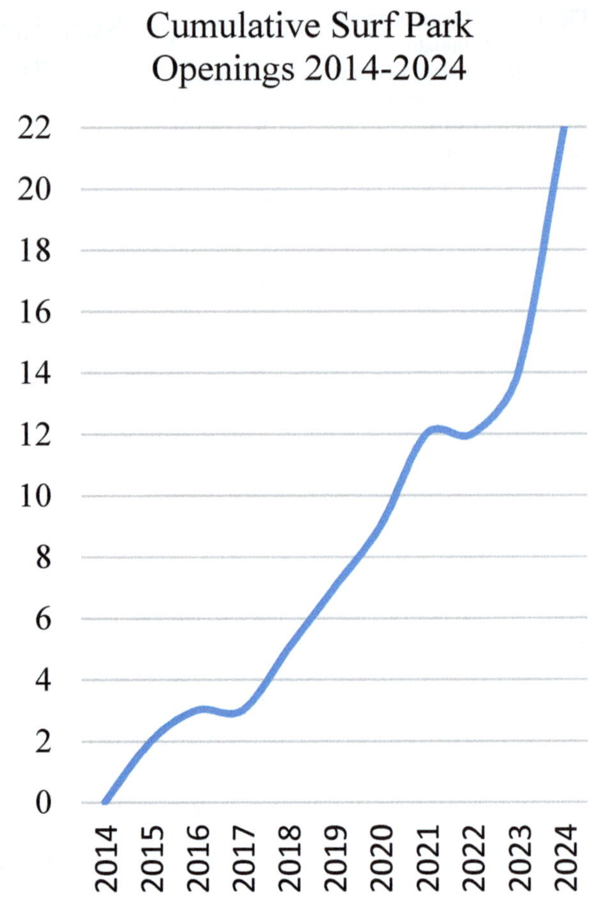

Fig. 1.3 Surf park openings, actual and cumulative 2014–2024. *Source* Ponting (2025) Presentation to Arizona Surf Park Developers Tour

notice. The Lost Shore Surf Resort in Edinburgh became the first surf park project to secure institutional capital through the Scottish National Investment Bank and Oaknorth Bank signaling a turning point: surf parks are now considered a legitimate real estate asset class.

Meanwhile, consumer interest has exploded. In 2015, only 78% of surfers expressed a willingness to visit a surf park; by 2023, that number had risen to over 99% (Ponting, 2023). Many enthusiasts are willing to travel significant distances-15% traveled by air and 29% drove for more than four hours to reach their surf park of choice—see Fig. 1.4.

These patterns vary widely based on location, pricing, and park design. While some, like The Wave, Bristol, draw 82% of guests from within a three-hour radius, others like Waco Surf see 64% of users traveling more than three hours, half of them flying in. Long-range travel is expected to decrease as more parks open closer to urban centers. With expanding consumer interest, better business models, increasing developer expertise, twenty-two dynamic wave surf parks in operation, scores more

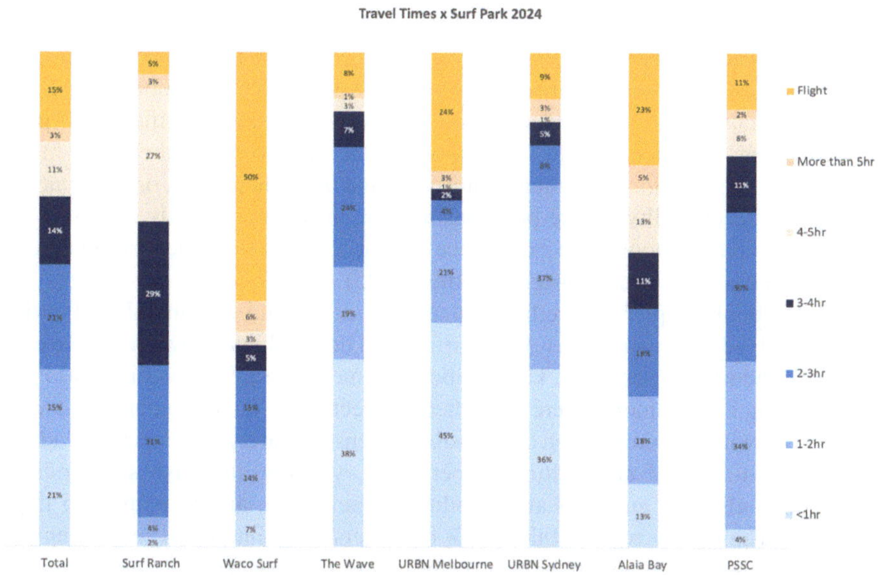

Fig. 1.4 Surf park travel times × Surf park. *Source* Ponting (2025) Presentation to Arizona Surf Park Developers Tour

currently in development, and improved access to capital, the surf park industry is accelerating rapidly. The next phase of growth will likely be defined not just by the number of parks built, but by the sophistication and scale of the developments surrounding them. What will all this mean for surfing? Academics have only just begun to ask.

Academic Literature Review

Human made surfing environments capable of being built virtually anywhere that can produce waves of extremely high quality that are economically viable leads to a host of impacts—socio-cultural, environmental and economic—both positive and negative, that are objectively fascinating to consider. Despite this, academic research on surf parks has been limited, perhaps because of a general lack of transparency across the industry as it finds its footing. The earliest published study concerning surf parks, Roberts and Ponting (2018), drew on theorists Jean Baudrillard (simulation and hyperreality) and Walter Benjamin (mechanical reproduction and loss of aura) to explore authenticity in surf culture in light of the advent of technologically sophisticated artificial surf lagoons. Baudrillard's notion of hyperreality emphasizes how simulated experiences may surpass or replace real experiences, blurring the line between reality and simulation, while Benjamin underscores how mechanical reproduction can remove the aura of authenticity tied to original, natural experiences. The authors concluded that the emergence of artificial surf parks has significantly altered debates around authenticity within surf culture. Surfing's traditional authenticity, tied to nature and oceanic spirituality, faces a strong challenge from commercialized, predictable, and simulated surfing experiences. The rise of artificial surf parks signals a new, contested era in surfing, where the question of authenticity hinges on whether one prioritizes spiritual connection and unpredictability, or convenience, safety, and accessibility. These positions are not, however, mutually exclusive and surfers are already choosing to experience both at different times depending on their personal needs and circumstances (Ponting, 2025).

A 2017 book chapter in an edited volume on surfing and sustainability (Ponting, 2017) involved a survey of over 3000 surfers and detailed interviews with five early surf park industry experts. Six key symbolic elements essential for successful artificial surfing environments were identified: perfect wave quality, guaranteed wave availability without competition or crowding, a sense of safety and controlled adventure, authentic social and cultural experiences, visually attractive surroundings, and accessibility (in terms of affordability, participation, and proximity to target markets). The survey revealed that 78% of surfers were interested in visiting surf parks. However, respondents who were hesitant cited concerns about a "spiritual divide," referencing the absence of the natural, spiritual connection experienced while surfing in the ocean. Additionally, they raised issues regarding environmental sustainability—specifically, the high consumption of water, energy, and land—as well as the cultural detachment and lack of authenticity stemming from the artificial nature of surf parks.

O'Brien (2021) utilized a qualitative dual-case study method to examine if and how two commercial surf parks in the UK contribute to community capacity building. Adventure Parc Snowdonia in Wales and The Wave Bristol were analyzed through 25 semi-structured interviews with various stakeholders supplemented by media reports. O'Brien found that surf parks positively influence their communities through job creation, vocational training, youth engagement, health promotion, and improved community pride. They successfully combined commercial objectives with social responsibility, fostering leadership, civic participation, and sustainability through structured dialogue, volunteering opportunities, inclusive programming, and strategic community partnerships. The study concluded that surf parks can exemplify how commercial entities can align business goals with community objectives to deliver long-term social and economic benefits.

Christophe Guibert at the Université d'Angers supervised Masters students, Yann Vilboux, Carlo Grigoletto and Jules Raimondo who explored the social and political dynamics surrounding wave pool development in France, with a focus on contested projects in Castets and Canéjan. Their detailed work appears in the next chapter of this book. The authors argue that despite international momentum behind surf parks, French projects have faced strong localized resistance, particularly from environmental NGOs like Surfrider Foundation Europe. Through qualitative case studies and interviews with stakeholders, the research highlights how wave pools have become flashpoints in broader debates over ecological sustainability, cultural identity, and urban development. Opponents frame surfing as an activity inherently tied to natural ocean environments, while developers argue for the social, economic, and training benefits of controlled, year-round wave access. The study reveals how political support, media framing, and public discourse have become critical arenas of contestation, showing that in France, the future of surfing—particularly in artificial environments—is deeply entangled with public values, environmental politics, and place-based identity. At the 2025 European Surf Park Industry Forum in Munich French wave technology company Okahina Wave claimed that their technology in development is viewed favorably by the French surf park resistance.

Byun and Choi's (2025) research paper, "Riding Artificial Waves: The Hybridisation of Surfing in Urban Spaces," investigates how artificial surf parks are reshaping surfing by enhancing accessibility, altering traditional surfing culture, and creating a hybrid surfing experience that combines traditional elements with urban convenience. The study finds that artificial wave pools significantly increase access to surfing for those living far from oceans and shift surfing culture toward performance and competition. However, the authors also note potential environmental concerns related to high water and energy usage and adverse effects on local communities due to overcrowding. Despite these issues, the paper acknowledges the benefits of artificial surf lagoons such as providing safer, controlled environments for learning and promoting surfing more broadly.

Several common themes emerge from the handful of academic analyses of surf parks. Scholars consistently highlight the tension between authenticity and simulation, noting that surf parks challenge traditional surfing's spiritual connection to the ocean by offering predictable, artificial experiences. Accessibility and convenience are significant drivers behind surf parks' increasing popularity, making surfing available to a broader demographic, including those far from natural surf breaks. Sustainability is another recurring concern, as studies highlight environmental impacts related to water and energy use. Additionally, researchers emphasize the positive role surf parks can play in community development, demonstrating how these commercial ventures can successfully integrate economic objectives with social responsibility and community engagement. Overall, existing research indicates surf parks are reshaping surfing culture into a hybrid form—melding commercial, accessible recreation with ongoing debates surrounding authenticity, environmental responsibility, and community benefits.

Imagining a future with hundreds of surf park installations around the world, perhaps thousands (for reference there are almost 40,000 golf courses in the world through 206 countries, with 16,000 courses in the US alone) forces us to imagine a future with many thousands, and perhaps millions of surfers who have learned to surf in a surf park. While the skills required to ride a wave can be accrued in a surf park, safely negotiating surfing lineups and their associated hazards in the ocean, understanding the nuanced sub-cultural 'rules' of unregulated surfing environments and identifying and selecting quality waves and positioning oneself effectively to catch that wave will all be unfamiliar challenges for the surf park surfer entering the ocean. The first elements of surf park culture to emerge, anecdotally, are more reflective of skate park culture. Skate park culture which tends to be far more inclusive and supportive than surf culture—the difference being the temporary, finite, and limited nature of quality waves over a defined period that frequently leads to intense competition between surfers. Skaters are more apt to take turns and support and celebrate each other for pushing their own limits, wherever those limits may be. Surf parks seem to elicit similar responses with learners being celebrated with similar levels of enthusiasm as competent surfers when visibly improving. This could be a long term win for surfer diversity.

The promise of large populations of 'new' surfers strikes terror into the hearts of many ocean surfers who already feel the pressure of rapidly intensifying crowds at their home breaks post-covid, when the surfing population grew by as much as 25% globally. It will be incumbent on surf park surf schools to educate their visitors about ocean surfing culture and norms to protect their customers from unpleasant and potentially violent encounters if/when they take their newfound hobby to the ocean. One counter intuitive outcome playing out is that established ocean surfers are replacing surf trips they would normally have taken to quality surf breaks in the ocean with trips to surf parks driven by guaranteed wave quality, wave count, absence of hassling, and the ability to book a session at a convenient time, even after dark. The 2023 Surf Park Consumer Trends Report (Ponting, 2023) found that 40.2% of surf park users had already substituted at least one surf park visit for a conventional surf trip in the past 12 months and that 39% planned to substitute a surf trip with a surf park visit in the next 12 months (including 48% of women—perhaps reflecting the more inclusive culture of surf parks alluded to above). It therefore seems likely that surf parks will increase and diversify the population of surfers in the ocean, and provide an alternative surf trip destination where the waves are guaranteed.

Perception Gaps and Water Use Misconceptions

Academic research to date suggests that surf parks offer significant community development, health, and economic benefits to their neighbors. Despite the benefits, they are often subject to criticism for their water use—particularly in drought-prone regions like California's Coachella Valley where four surf parks have been planned. One failed to gain approval, one is complete, and another is due to open in 2026. Comedian John Oliver (2022) called surf park developments in Coachella Valley "monumentally stupid", compared them to a wall of constantly flushing toilets, and performed hyperbolic outrage at the idea that water would be wasted on something so frivolous as surfing. Oliver, however, failed to explore or compare the water use of surf parks relative to the golf courses they provide development alternatives to for mixed use real estate developments in the Coachella Valley.

The average Coachella Valley golf course uses 277 million gallons of water annually (Coachella Valley Water District, 2025) and there are 97 in the Coachella Valley which use a total of 25.2 billion gallons of water annually. By contrast, a maximum of four surf parks were planned for the region. One (Coral Mountain, a planned KSWaveCo facility) was rejected by the community for excessive water use. The Coral Mountain development planned to use just 14% of the water of adjacent golf courses. However, once the surf park project was abandoned because of its excessive water use it was replaced by a golf course which will use 238 million gallons of water more than the proposed surf park would have each year. See Table 1.2 for comparisons between an average Coachella Valley golf course and different surf park technologies operating in the same region.

Notably, DSRT Surf offsets its water use by replacing over 1 million square feet of turf with xeriscaping, saving an estimated 34 million gallons annually and resulting in a net water savings (Terra Nova Planning and Research, 2019). As the research suggests, this kind of innovation in sustainability practices is necessary for surf parks to win public support. It also meets market demand for sustainability in surf parks.

Figure 1.5, shown, presents a global respondent heat map derived from a 2024 survey conducted by the Surf Park Industry Association and presented at Surf Park Summit on October 1st, 2024, in La Jolla, California. The survey captured responses from 4310 surfers—predominantly from the United States, Australia, and the United Kingdom—who were asked to respond to a series of sustainability-focused statements. Figure 1.6, also drawn from this dataset, illustrates the levels of agreement and disagreement across various sustainability initiatives.

The results provide revealing insights into the surf community's expectations around surf park development. Across ten different sustainability initiatives, the data show moderate to strong support, with most categories earning a combined total of 60–75% agreement (combining "Agree" and "Strongly Agree" responses). A clear, consistent, and large majority of surfers are in favor of integrating sustainability into the planning and operation of surf parks. The statements used in the survey have been abbreviated to fit into Fig. 1.4 but took the form of positive statements about each topic. For example, in the first instance the statement used was "Surf parks should provide access to and programming for people with disabilities", in the graph this was abbreviated to "Adaptive surfing access and programming".

The top three priorities, based on the highest levels of agreement, reflect a notable emphasis on both social and environmental responsibility. Leading the list is adaptive

Table 1.2 Coachella Valley annual water use. Golf versus surf parks*

Facility	Estimated annual water use (Million Gallons)	% of 18 hole golf course	Holes of golf equivalent
Average Coachella Golf Course	277	100%	18
Palm Springs Surf Club (Surf Loch)	3	1.1%	0.2 holes
Endless Surf ES34 (Coachella)	15.75	5.7%	1 hole
DSRT Surf (Wavegarden Cove 52)	23	8.3%	1.1 holes
Coral Mountain (KSWaveCo)	39	14%	2.5 holes

* Data provided by wave generation technologies and developers by direct request of the author

Fig. 1.5 Heat map of responses to 2024 surf park survey. *Source* Ponting (2024)

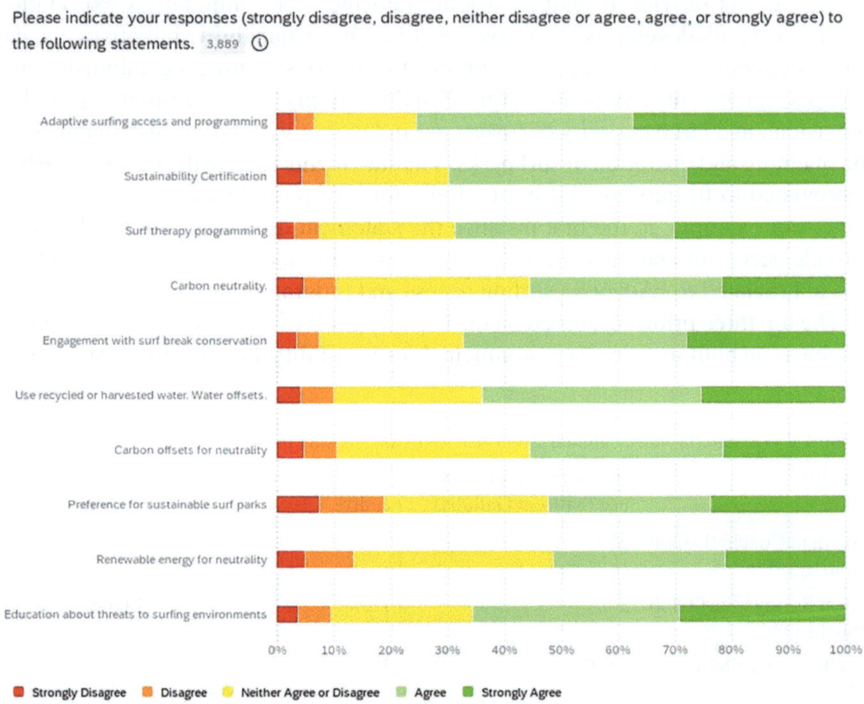

Fig. 1.6 Surf park user opinions on sustainability measures. Source: Ponting (2024)

surfing access and programming, with around 75% of respondents supporting the initiative on only 6% of respondents disagreeing that surf parks should provide these types of opportunities and programing. This high level of agreement underscores a powerful expectation that surf parks play a role in making surfing accessible to people with disabilities and those who have traditionally been excluded from ocean-based surf culture. Surf therapy programming had similarly strong backing of 69% with 6% disagreeing that opportunities should be provided for surf therapy organizations. This indicates growing awareness of surfing's mental health benefits and a belief that surf parks can serve therapeutic as well as recreational purposes. Rounding out the top three is sustainability certification, which received 70% agreement with only 8% of respondents disagreeing that "surf parks should be in compliance with a relevant sustainability standard"—suggesting that surfers value formal, verifiable accountability when it comes to environmental and social claims made by surf park developers.

Education about threats to surfing environments had 66% agreement (9% disagreement), and providing opportunities to engage with surf break ecosystem conservation enjoyed 67% support with only 7% disagreement. Several other categories also garnered strong support, though with slightly more nuanced sentiment. Carbon neutrality, renewable energy strategies, and the use of recycled or harvested water all hover around the 70% mark in total agreement. However, these categories exhibit a wider spread across the full range of responses, including a significant number of "Neutral" responses. This distribution may indicate a general endorsement of the concepts, but also reveal a gap in understanding about how these goals are achieved—or skepticism about their effectiveness in practice. Carbon offsets for neutrality showed slightly lower levels of strong agreement, aligning with broader environmental debates over the integrity and transparency of offset schemes.

Each of the 10 sustainability measures tested secured at least 52% agreement and strong agreement, and 8 of the 10 measures drew less than 10% disagreement. The implications are clear. Sustainability is now a baseline expectation—particularly around inclusive access, therapeutic programming, and third-party certification. Initiatives related to carbon neutrality and water conservation also matter, but developers should not assume that public support translates into universal understanding. Clear, credible communication around how sustainability goals are achieved will be essential. The data also point to an opportunity: programs that build community and offer tangible social impact—like adaptive surfing or surf therapy—may generate more trust and goodwill than abstract carbon commitments. Similarly, strong support for environmental education around threats to surf ecosystems in the ocean, and access to and engagement with surf break ecosystem conservation, provides an opportunity for surf parks to participate in programs with real-world tangible benefits that simultaneously create authenticity around their sustainability messaging and foster community among surf park users and the wider surfing. To earn long-term loyalty, surf parks will need to demonstrate both authenticity and measurable impact.

Likely Impacts by 2040

Based on surf park openings and the likely growth trajectory of the industry indicated in Figs. 1.2 and 1.3, combined with increasing access to capital, evolving business plans, solidifying legitimate asset class status, and growing numbers of wave technologies and multi-park development groups it seems likely that on average the industry will very conservatively add 10+ surf parks per year through to 2040. At this rate the industry will more than double in terms of operating parks in three years, and double again in the following five years. Based on this growth trajectory we can conservatively expect at least 100 surf parks to be operating around the world by 2033 and between 170 and 200 by 2040. Clearly this level of development comes with very significant levels of investment (tens of billions), employment, recreational opportunity, and economic activity for surf park destination\s. But what does this mean in terms of people being exposed to surfing and 'surf culture'? How many surf hours are likely to be utilized and how many people are likely to visit surf parks?

Precise figures of surf park utilization are difficult to access as they are considered proprietary knowledge by most operators and most jurisdictions do not compel this kind of reporting (the UK being an exception). Despite this, URBNSURF Melbourne claimed that in their first five years of operation 10 million waves were produced and surfed by 300,000 guests over that period—totaling 33.3 waves per guest, so these waves were likely accrued over three surf sessions. The number of non-surfing guests was not revealed. Dividing by five to estimate one year of operation we get 2,000,000 waves per year serving 60,000 surfing guests over 180,000 surf hours. If we extrapolate assuming 200 surf parks by 2040 this equates to 400,000,000 waves being generated across 200 surf parks serving 12,000,000 surfing guests surfing for 36,000,000 h.

Public financial reporting requirements in the UK compel The Wave, Bristol to reveal its utilization from its opening in 2020 which was plagued by Covid lockdowns and closures, to a doubling of utilization in 2021, and some correction through to 2023. See Table 1.3.

The average number of surf hours at The Wave, Bristol in non-covid years has been 149,333. Each surf hour consists of 10–12 waves so conservatively that equates to 1,493,000 waves per year. The average number of non-surfing visits per non-covid year has been 46,333. If we extrapolate these numbers out to 200 surf parks, that equates to 29,866,600 surf hours (approximately 300 million waves) and 9,266,600

Table 1.3 The wave, Bristol surf hours sold and non-surf visitors 2020–2023

Year	Surf hours sold	Non-surfing guests
2020	75,000	37,000
2021	157,000	55,000
2022	149,000	40,000
2023	142,000	44,000

Source UK Government Companies House

Table 1.4 Estimate of global surf park utilization by 2040

Utilization data from	Number of waves	Number of surf hours	Non-surfing visitors	Total number of visitors
URBNSURF	400,000,000	40,000,000	13,300,000	53,300,000
The wave	300,000,000	30,000,000	10,000,000	40,000,000
Average	350,000,000	35,000,000	11,650,000	46,650,000

non surfing visitors, or roughly 40,000,000 surf park visits in a ratio of surfers to non-surfers of roughly 3:1.

If we assume 10 waves per surf hour and take The Wave's 3:1 ratio and apply it to URBNSURF's utilization data we get the projections outlined in Table 1.4.

It should be noted that Wavegarden Cove (the technology used at both URBNSURF and The Wave, Bristol) surfer capacities are higher than most other technologies and that both surf parks utilize relatively simple, open-to-the-public business models driven by surf hours (with related lessons, rentals, photography and videography) with simple food and beverage and retail businesses in support. Evolving business models and increasingly sophisticated mixed-use developments are likely to increase the number of non-surfing visitors in future developments to the extent that the surf hours to non-surfing visitor ratio may be reversed, leading to vastly increased numbers of visitors spending time and money on things other than surf time.

Could Surf Parks Fund the Protection of Surf Ecosystems?

Given the overwhelming support among surf park users to create linkages with ocean surf ecosystems and provide information, education, and opportunities to participate in the conservation of surf ecosystems, it is interesting to speculate on the potential impact of the industry on surf ecosystem conservation.

The potential has not gone unnoticed by the industry. STOKE, a non-profit organization housed in San Diego State University's Center for Surf Research and the leading provider of sector specific sustainability certifications for surf parks, is actively working with Conservation International's Surf Conservation Program to develop a "Surf Conservation Champion" badge within its larger surf park certification program. Conservation International has mapped the overlap between areas of high terrestrial biodiversity and conservation value, areas of high marine biodiversity and conservation value, ecosystem storing high levels of embodied carbon, and high-quality surf breaks. Many of the areas at the center of that Venn diagram are under significant threat of destruction and efforts are in place to conserve these ecosystems with sustainable surf tourism as an economic incentive and funding mechanism. Surf parks that meet specific criteria related to surf ecosystem conservation education and support while also meeting minimum compliance for STOKE certification will be afforded a Surf Conservation Champion badge to recognize their efforts. Key among these efforts are opportunities for surf park users to contribute to surf ecosystem conservation.

Ponting's, 2024 surf park survey asked respondents to "Please indicate how much more (as a percentage) you would be willing to pay to visit a surf park that can demonstrate a commitment to environmental, social/cultural and economic sustainability?" The results across different demographic groups in the dataset as well as all results together are shown in Fig. 1.7 along with the responses discounted by 21% to allow for hypothetical reporting bias (Schmidt & Bijmolt, 2020).

Figure 1.7 indicates a clear willingness among visitors to financially support sustainability initiatives. After adjusting for hypothetical bias, the data suggest that the average visitor is willing to pay an additional 13.4% for a surf park experience that can demonstrate authentic commitments to environmental, social, and economic sustainability. This is not a marginal figure—it reflects an enormous potential revenue stream for sustainability-related investment.

Spending data reveal that typical surf park visitors spend between \$250–\$300 on surf-related activities (e.g. wave sessions, lessons, equipment rentals) and an additional \$200–\$300 on ancillary expenditures such as food, beverage, retail, and accommodation (Ponting, 2024). Using the conservative lower bounds of these ranges (\$250 + \$200), the average total spend per visit is estimated at \$450. Applying the corrected 13.4% premium to this amount yields an additional \$60.30 per visitor that could, in theory, be directed toward sustainability programs. Even when using just the \$200 non-surfing expenditure figure, the 13.4% premium would still generate \$26.80 per person.

To contextualize this potential, we can apply these figures to known visitation numbers. In 2023, The Wave in Bristol reported 44,000 non-surfing visitors. If each of those visitors had contributed just \$26.80 toward sustainability initiatives, the

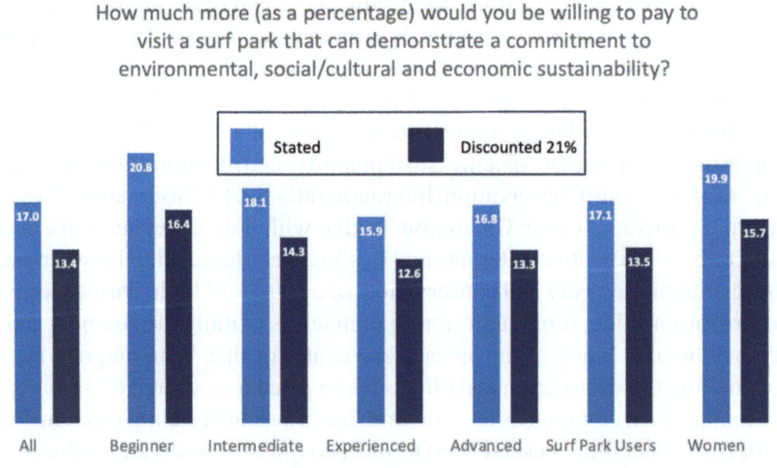

Fig. 1.7 Surf park visitors' willingness to pay for sustainability. *Source* Ponting (2024)

resulting revenue would have totaled approximately $1.18 million—without any increase in attendance or surf sessions. Scaled to the global level using conservative 2040 projections of 200 surf parks yielding 46.65 million total visits, this same mechanism could generate over $1.23 billion in sustainability-directed funding if average visitor spending and willingness-to-pay levels hold constant.

Even more modest contributions offer compelling outcomes at the likely scale of the surf park industry in 2040. For instance, if surf parks worldwide implemented a $1 per-visitor surcharge earmarked exclusively for surf ecosystem conservation, the total potential annual contribution—based on the 2040 visitation estimates—would be $46.65 million. Alternatively, a 10-cent per-wave fee (applied to the estimated 350 million waves that will be generated annually across all parks by 2040) would produce $35 million annually. For perspective, had such a fee been in place at The Wave, Bristol in 2023, it would have yielded $142,000 that year alone. See Table 1.5 for industry-wide and individual surf park potential economic impact for sustainability initiatives.

With the number of visitors, the number of waves, and the reported willingness of those visitors to pay more for sustainability in surf parks the numbers become significant very quickly. Applying these types of fees, levies, taxes, point of sale round ups, donation opportunities at nominal levels for each visitor could enable surf parks to aggressively pursue their own in-house sustainability innovations (perhaps including adaptive surfing and surf therapy programming which ranked highly among potential initiatives) and incorporate them into their branding and messaging as well as potentially supporting sustainability initiatives related to surfing beyond the surf park gates. The adoption of these measures industry wide even at the level of 0.5–1% of total spend per visitor has the potential to be highly impactful.

Table 1.5 Potential contributions to sustainability in 2040

Contribution level	Industry wide	Individual surf park
13.4% of $450 spend = $60.30	$2.713 Billion	$13.6 Million
13.4% of $200 spend = $26.80	$1.206 Billion	$6.03 Million
$10 per visit	$450 Million	$2.25 Million
$1 per wave ridden (4% of $200 spend)	$350 Million	$1.75 Million
$5 per visit (2.5% of $200 spend)	$225 Million	$1.125 Million
$2 per visit (1% of $200 spend)	$90 Million	$450,000
$1 per visit + $0.10 per wave ridden	$80 Million	$400,000
$1 per visit (0.5% of $200 spend)	$45 Million	$225,000
$0.10/wave (0.05% of $200 spend)	$35 Million	$175,000

A Scalable, Sector-Wide Conservation Mechanism

In addition to investing in internal sustainability performance—such as renewable energy systems, water conservation infrastructure, and accessibility programs—surf parks have a powerful opportunity to become drivers of external environmental impact. Specifically, they can serve as funding engines for surf break conservation and the protection of marine and coastal ecosystems. The economics are compelling: the scale of the industry means that small per-unit charges (whether per wave, per hour, or per visit) can translate into millions of dollars annually.

A sector-wide initiative that standardizes a nominal sustainability fee—such as a 10-cent-per-wave levy or $1-per-visit surcharge—could create a stable and meaningful funding stream for organizations working in surf ecosystem preservation, including Conservation International's Surf Conservation Program, Save the Waves Coalition, and the Surfrider Foundation. These funds could be used not only for conservation activities, but also for public education, local community engagement, sustainability initiatives on surf park premises, and stewardship of high-value surf breaks under threat from coastal development, pollution, or climate change.

Importantly, this approach aligns with consumer expectations. As documented earlier in the chapter, over 70% of survey respondents expressed support for surf parks that meet third-party sustainability standards, and 67% endorsed the integration of surf ecosystem education and conservation into park programming and on average surf park visitors are willing to pay 13.4% for sustainable operations. This indicates an emerging cultural norm: sustainability is no longer seen as a competitive advantage for surf parks—it is becoming a baseline expectation.

In sum, the surf park industry has a unique opportunity to establish a position at the intersection of recreation, real estate, and conservation finance. By leveraging minimal visitor contributions well below levels that surf park visitors report being willing to pay, the sector can catalyze substantial environmental benefits, enhance its social license to operate, and meet the demands of a rapidly evolving consumer base—with zero impact on commercial viability.

The Quiet Threat to a Sustainable Future

While there is considerable reason for optimism regarding the surf park industry's capacity to evolve sustainably—especially given the regulatory and market pressures surrounding energy efficiency, water conservation, and community engagement—the future is not guaranteed. In many jurisdictions, entitlements and approvals for surf park development are contingent upon demonstrable commitments to sustainability. These include strategies such as the use of recycled or harvested water, renewable energy adoption, and community-facing programming, and accessibility initiatives. Encouragingly, numerous operating and in-development parks are pursuing or have already achieved third-party certifications such as STOKE, signaling a degree of accountability within the sector.

However, the presence of these incentives does not preclude the possibility of backsliding—particularly as the surf park business model becomes increasingly entangled with private real estate development. A growing number of projects are structured as large-scale greenfield residential developments, with the surf park serving as an exclusive recreational amenity rather than a public-facing facility. These integrated models appear to be highly profitable and more easily fundable, and they may have less exposure to community-oriented dynamics that encourage strong sustainability performance.

In such developments, there is typically no immediately adjacent existing residential community whose support must be earned, nor is there a broader public clientele holding operators accountable through market demand for transparency and environmental or social value. As a result, developers may deprioritize investments in inclusivity programs, sustainability education, or energy and water innovation—particularly if these initiatives are viewed as cost centers rather than market differentiators. In these cases, the usual levers of social license and reputational pressure are likely to be weakened. Without the checks and balances provided by a critical and participatory public, the sustainability performance of such projects becomes dependent on the goodwill and long-term vision of developers—an inherently fragile arrangement though it has held for the time being.

Therefore, while regulatory frameworks and public expectations can drive innovation and accountability in surf park developments, these pressures are not uniformly distributed across all project types. If the sector trends toward exclusivity and private recreation, it may lose the momentum currently building around its potential contributions to accessibility, environmental awareness, and social cohesion. The surf park industry must remain vigilant in guarding against this drift, and stakeholders—from certifying bodies to investors and policymakers—must play an active role in reinforcing standards, especially in contexts where market incentives for sustainability are weakest.

Conclusion

The modern surf park industry has suffered setbacks, from the Ocean Dome closure in 2007 to the Ron Jon Surf Park failure of 2008, and the closing of both Wavegarden Lagoon installations after just 8 and 2 years of operation. All of these closures were a function of technology that was either not up to the challenge, too expensive to facilitate profitability, or not successfully positioned within their markets. All the wave generation technologies commercially operating today are entirely capable of developing operationally reliable surf lagoons that create very high-quality waves at economically sustainable rates given the right business model in the right geography with the appropriate market placement. For example, high throughput technologies may be best suited to urban settings with lower price points drawing large numbers of visitors onsite for relatively shorter periods of time. Others with lower throughputs are positioning themselves as overnight surf destinations. With each new installation incremental improvements are made to wave basin design, and software improvements to wave type and quality can be rolled out to all previous installations online.

Current commercial wave generation technologies use either chambers that move water using air pressure or mechanical means, an electro-hydraulic central water displacement device, paddles that move water, or moving foils. New technologies including submerged pneumatic pillows and floating circular modules in existing bodies of water hint at paradigmatic change through better performing technology. The future may see technology that dramatically lowers capital and operational costs for developers and operators with lower environmental impact and potentially even higher quality waves. Innovations around energy are already in the planning. For example, one planned project is due to be entirely powered by an onsite solar array of a scale sufficient to also power an onsite data center, the heat from which would be used to heat the water in the surf lagoon.

If the very conservative projections developed in this chapter are realized, a surf park industry with 200 operating facilities may accommodate over 46 million visitors annually by 2040, generating hundreds of millions of artificial waves and contributing billions of dollars in economic activity. This would represent a profound cultural transformation: surfing, historically rooted in ocean-based experiences and coastal geographies would to some extent be redefined by inland, technologically mediated environments. The extent to which this transformation enhances or undermines the cultural integrity of surfing will depend on decisions made in the present—particularly concerning access, governance, environmental stewardship, and cultural framing for new surfing participants.

The long-term significance of surf parks lies not solely in their economic or technological achievements, but in their potential to expand inclusivity, enhance the physical and mental wellbeing, create community, promote environmental literacy, and support conservation financing for natural surf ecosystems. However, there is also a plausible risk that commercial imperatives will eclipse social and environmental considerations, resulting in developments that prioritize profit over principle and deliver spiritually hollow commodified simulations rather than meaningful engagement with surfing's cultural and ecological roots.

Ultimately, the trajectory of the surf park industry will be shaped by the values, incentives, and decisions of developers, operators, policymakers, and communities. The opportunity to guide this evolution toward more sustainable, inclusive, and culturally grounded outcomes remains open—but it will require intentional effort and accountability.

References

Byun, J., & Choi, K. H. (2025). Riding artificial waves: the hybridisation of surfing in urban spaces. *Leisure Studies*, 1–17. https://doi.org/10.1080/02614367.2025.2455599

Coachella Valley Water District. (2025). https://www.cvwd.org/

Lowry, T. (1985). His money comes in waves. *Sunday Call-Chronicle*. Retrieved from http://articles.mcall.com/1985-06-23/news/2465673_1_waves-pools-oil-rigs

National Golf Foundation. (2025). Golf Industry Research https://www.ngf.org/the-clubhouse/golf-industry-research/

O'Brien, D. (2021). Commercial sport organisations and community capacity building: A case study of two surf parks. *Sport Management Review, 24*(5), 723–746. https://doi.org/10.1080/14413523.2021.1902143

Oliver, J. (Host). (2022, August 7). *Water: Last week tonight with John Oliver (HBO)* [Video]. YouTube. https://www.youtube.com/watch?v=jtxew5XUVbQ

Ponting, J. (2017). Simulating Nirvana: Surf parks, surfing spaces, and sustainability. In Borne, G., & Ponting J. (Eds.) *Surfing and Sustainability*. Routledge.

Ponting, J. (2023). Surf park consumer trends report, surf park central, Solana Beach, CA.

Ponting, J. (2024, September 30–October 2). *Opening remarks*. Presented at Surf Park Summit, La Jolla, CA.

Ponting, J. (2025). Are surf parks eating conventional surf tourism's lunch. *Surf Park Central*. https://surfparkcentral.com/are-surf-parks-eating-traditional-surf-tourisms-lunch/

Roberts, M., & Ponting, J. (2018). Waves of simulation: Arguing authenticity in an era of surfing the hyperreal. *International Review for the Sociology of Sport, 55*(2), 229–245. https://doi.org/10.1177/1012690218791997

Schmidt, J., & Bijmolt, T. H. A. (2020). Accurately measuring willingness to pay for consumer goods: A meta-analysis of the hypothetical bias. *Journal of the Academy of Marketing Science, 48*, 499–518. https://doi.org/10.1007/s11747-019-00666-6

Temperley, E. (2015). The wave garden at Snowdonia will open in July. *Magic Seaweed*. Retrieved from http://magicseaweed.com/news/the-wavegarden-at-surf-snowdonia-will-open-in-july/7154/

Terra Nova Planning & Research. (2019). DSRT surf specific plan environmental impact report. Appendix 1 Water Supply Assessment and Water Supply Verification for the DSRT SURF Project, Report prepared for the City of Palm Desert https://www.palmdesert.gov/home/showpublisheddocument/27559/637088011695670000

Vilboux, Y., Grigoletto, C., & Raimondo, J. (2024). *Wave pools in France: Localized controversies* (Master's thesis, Université d'Angers).

Warshaw, M. (2004). *The encyclopaedia of surfing*. Penguin Books.

Waveloch. (2021). Waveloch history. http://www.waveloch.com/history/

Weinstein, M. (2025) *Commercial real estate uptake at cannon beach*, Presented at the Arizona Surf Park Development Tour, March 19th–20th, Mesa Arizona.

Westwick, P., & Neushul, P. (2013). *The world in the curl: An unconventional history of surfing*. Crown Publishers.

Wilson, J. (2008). That's bullshit. *Surfling Magazine*. Retrieved from http://www.surfingmagazine.com/news/thats-bullshit-chapter-1-jimmy-wilson-082208/-640UZvkA12bw1P7P.97

Chapter 2
Wave Pools in France: Controversies Within Localised Political Fields

Christophe Guibert, Carlo Grigoletto, Jules Raimondo, and Yann Vilboux

Introduction

"The beginning of the twenty-first century has seen the rise of a whole nebula of commitments combining concern for the living world, criticism of the logic of commodification, the quest for autonomy and identification with places" (Tournadre, 2024). (In France, the opposition to various developments and projects are backed by political structures and associations on the one hand, and more informal movements on the other. But it is unclear how effective the mechanisms for raising ecological awareness are at changing the current political structures.) Water retention basins in rural areas for agricultural purposes, construction of motorway stretches, plans for marinas on the coast, building and development of golf courses and plans to build wave pools for surfing are just some of the wide range of facilities that, while not unprecedented in France, crystallise economic, tourist, social, partisan and ideological viewpoints. But perhaps even more than these, the environmental aspect seems to be mobilised even more, due to the image of surfing conveyed in French society, a practice characterised by numerous spontaneous representations (Guibert, 2020). Surfing and the practice of surfing are only what individuals make of them. Indeed, surfing has no nature, it has no essence: it is the product of long cosmopolitan, historical and social processes. Surfing could be viewed as a non-organized 'leisure' activity, a competitive professional sport, a non-professional sport, a 'way of life', a beach game, an economic activity, a counter-cultural activity, or a specific and dedicated relationship with 'nature', (the list is obviously not exhaustive). Surfing

C. Guibert (✉) · J. Raimondo
University of Angers, ESO Laboratory (UMR CNRS 6590), Angers, France
e-mail: christophe.guibert@univ-angers.fr

J. Raimondo
e-mail: jules.raimondo@etud.univ-angers.fr

C. Grigoletto · Y. Vilboux
University of Angers, Angers, France

L. Mach (ed.), *The Future of Surfing in the Anthropocene*,
https://doi.org/10.1007/978-3-032-09377-6_2

can be all these things at once. In the light of our conceptual systems, which are in part our own social representations, each of us sees surfing as having one or more of these associations of ideas.

While these representations are based on our subjectivities, but also on very real, factual data (the existence of a French federation, summer surf schools on the beaches, champions with media coverage, "surf" destinations around the world, etc.), there is one idea that remains tenacious about the world of surfing in France and beyond. That is the idea that surfing has a "counter-cultural" dimension. By definition, counter-culture is associated with protest movements, the rejection of orthodoxies and a break with the dominant political and cultural orders. The social history of surfing in France shows that, from the outset, the activity was socially classifying and distinctive: it was practised by an elite who codified it symbolically (Guibert, 2007). What's more, if the "counter-culture" is so present in the discursive space and advertising images associated with surfing, it's because the argument is economically attractive, both for the specialised press and for international surfwear companies, who see it as a privileged means of distinguishing themselves from the "ordinary" competitive sport practised in clubs. This is where the interest of this paper lies, due to the impacts of wave pools on many aspects of the surfing practices.

Wave pools (or artificial waves), in particular—the subject of analysis in this chapter—are specific and highly innovative cases in terms of research. If surfing is often, in spontaneous representations (Guibert, 2020), intuitively associated with the practice of sport in a natural environment, within which the relationship with 'nature' (Audinet et al., 2017) and ecology (two notions understood here as categories of political action) are concomitant with involvement in this sport, how then can we interpret the practice of surfing in a developed, regulated and controlled space such as a wave pool? The 'marine sense' that the specialised press regularly refers to (knowing how to identify the best places to practice—the spots—subtly understanding the direction and strength of the winds and the effects of the tides on the quality of the waves, etc.), an aptitude that characterises surfing in the ocean, is no longer relevant in a wave pool, a standardised facility in which the height, power, shape and frequency of a wave can be meticulously controlled. Could this type of equipment be contributing to a kind of denaturing of an activity that is essentially practised in a natural environment? As surfing becomes an Olympic sport, a process already underway since 2021 at the Tokyo Olympic Games and then in Tahiti at the Paris 2024 Games, the athletic rationalisation of athletes in many countries (Brazil and the United States in particular) is based on training and the quest for technical performance in wave pools. According to Michel Plateau, France's national technical director, these pools allow "the repetition of movement, the ability to receive quick feedback from the coach."

Wave pools, whose technologies are highly scalable in the short term, are proliferating in diverse countries. Projects are now gathering pace with many scheduled to open in 2024 including in Tenerife (Spain), Scotland, Germany, Abu Dhabi, Brazil, Belgium, the Netherlands, not forgetting Arizona and Hawaii in the United States. The spread of these projects is contributing to the legitimacy (economic and sporting in particular) of the technology. Despite nearly a dozen projects that have been planned and advocated for by their promoters, in some cases with the support of local authorities, no wave pools have yet to be constructed in France. More concretely—and this is where the problematisation of this chapter lies—the controversies and tensions surrounding the numerous proposed wave pool projects in France in recent years are part of localised and territorialised political fields (Bourdieu, 1981, 2000) where the interests of some are opposed, in a continuum of values and representations, to the positions adopted by others.

The methodological approach is based on systematic documentary monitoring, an international and multidisciplinary literature review, and a historical and contemporary contextualisation of a reasoned and relevant selection of wave pool projects in France. More specifically, we considered communications from supporters and opponents, blogs, websites and public statements in the press. A qualitative approach is favoured to grasp and study in detail the positions and positions adopted in the social micro-spaces taken as subjects. A dozen semi-directive interviews were conducted with local elected representatives, association leaders, residents' groups, wave pool project promoters, etc. Observation phases were conducted within groups of supporters and opponents of the selected projects (including participation in several public information meetings).

A Multi-faceted Issue that Intersects with Multiple Local Challenges

A number of preliminary hypotheses can be put forward in an attempt to construct an analytical framework for this research: the influence of environmental associations (primarily *Surfrider Foundation Europe* and its local branches) and the political reluctance to construct a facility that consumes significant amounts of energy and resources (water in particular). Ultimately, how is the process of building a sports facility influenced by power relations beyond the field of sport: economic, legal, media, ecological and, finally, symbolic issues? Beyond these fairly generic hypotheses, it is more the analysis of interest groups (Saurugger, 2019), local mobilisations and the interpretation of the resources (in the sociological sense) available to the social agents involved in local struggles (both supporters and opponents) that is of interest here. Social capital, cultural capital (legal, environmental, economic knowledge, etc.), militant capital (Matonti & Poupeau, 2004), etc. are all individual and collective properties at the root of the struggles. In other words, who are the social agents involved? How are their positions constructed? Which normative or performative discursive regimes are valued and legitimised, and according to which social and political mechanisms? How do the 'players' in the world of surfing position themselves within these spaces of struggle? What visions and interests are involved in their ranking systems: whether it be the French Surfing Federation, surfwear brands and EuroSIMA (European brand cluster), specialist journalists (privileged relays of positions taken in the print or digital press) but also the surfers themselves, in particular those with an outstanding reputation, often acquired in competition, enabling them to expose their points of view to as many people as possible.

Surfing is a sport that plays a key role in structuring the tourist economy of the Aquitaine region, from the local to the regional level (Guibert, 2006). Surfing in Europe is most developed in the Nouvelle-Aquitaine region, particularly in the Gironde (Lacanau), Landes (Hossegor, Seignosse…) and Pyrénées-Atlantiques (Biarritz, Guéthary, Anglet…) departments. The headquarters of the French Surfing Federation (FFS), a dense network of associations (accounting for almost 80% of national licences), the majority of commercial businesses (schools, specialist shops, board manufacturers, shapers) and the headquarters of surfwear companies all contribute to the Aquitaine coast being described—by the business press and the Bayonne-Pays Basque Chamber of Commerce and Industry—as “Surficon Valley”. Germane to the critique of wave pools offered in this chapter, two proposed projects in Castets and Canéjan (Map 1) illustrate the difficulties faced by wave pool promoters. These cases offer the opportunity to conduct a micro-sociological analysis of local issues and political forces at play that accelerate or, conversely, slow down the success of these projects: as Subra (2016) says “conflict is not just the product of a project: it arises from the encounter between a project and a territory. As much as the characteristics of the project, it is the characteristics of the area that are the cause of the conflict: its history, its sociology, the culture of the local society that has developed there, and the value attributed to it (landscape, heritage or biodiversity value)” (Subra, 2016).

Map 1 Location of the Castets and Canéjan wave pool projects.
Both projects are located in the Nouvelle-Aquitaine region, close to France’s best “ocean” and “natural” surf spots (Lacanau for Canéjan and Hossegor and Biarritz for Castets, further south).
(Source: Google Map, 2024; Production: Guibert, 2024)

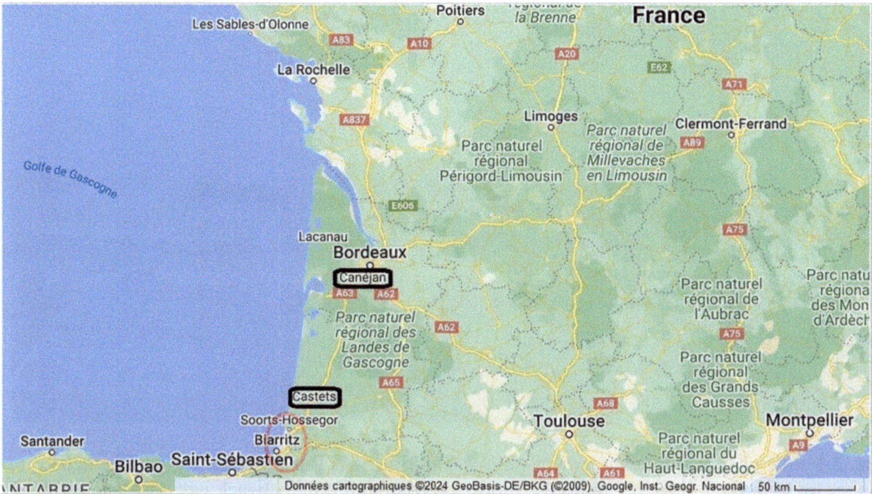

OR
(Production: Raimondo, 2024)

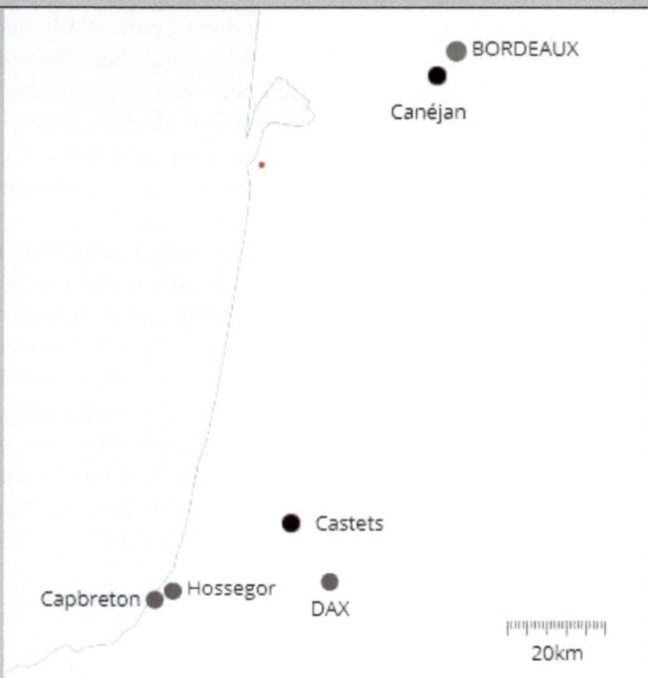

Castets is located in the south of the Landes department 25 kms from the towns of Hossegor, Capbreton and Seignosse, which are renowned for offering some of the

Fig. 2.1 Sketch of the Canéjan project. *Source* **Canéjan Surf Park.** In addition to the practice area in a swimming pool, there are plans for shops and restaurants and a beach with vegetation

best beach breaks in France and even Europe. The area is characterised by a high concentration of companies linked to the surfing industry, particularly the European Surf Industry Manufacturers Association (EuroSIMA), and "benefits from a strong history and legitimacy (cradle of French and European surfing), as well as proximity to a practice area with unanimously recognised qualities" (Dehez, 2020, p. 242). The wave pool in Castets is proposed to be built 20 kms from the Atlantic Ocean and is planned to cover an area of ten hectares, including tourist accommodation, practice areas for sporting and cultural activities, and commercial facilities.[1]

Canéjan, the second case study, is a municipality located around ten kilometres south of Bordeaux and 60 kms from the ocean. However, it is not part of the Bordeaux metropolitan area and therefore does not benefit from its public transport network. The wave pool project, based on technology developed by the Spanish company *Wavegarden*, aims to be built on a 3.6 hectare site of an old IBM factory that closed in 2007 and is close to a busy motorway. The 20,000 m2 complex will house two surfing pools, 160 and 190 m long respectively.[2] Conceived by four surfers, but financially supported by one of their fathers of one of them (a former contestant on a well-known French reality TV show), who is a wealthy property entrepreneur, the Canéjan artificial wave project aims to open in 2025. (Fig. 2.1).

[1] Website: https://www.surfsession.com/actu/wavelandes-atlantique-debut-travaux-janvier-2019-807184538/.

[2] *Sud Ouest*, « Une piscine à vagues artificielles en projet», 30/05/2023.

A Variety of Arguments and Support from the World of Sport

According to their promoters, these wave pools would be socially useful in several ~~a~~ ways, particularly in training for the Olympic Games, but also in terms of making sport accessible to "as many people as possible". Promoters also claim the projects would benefit the structure and economic development of the regions concerned. Competitions and training sessions, where all the participants have the "same" wave shape all the time, allow surfers to practise manoeuvres in a controlled environment. Manufactured pools permit "*competitors to train in predictable waves that can make it easier to master manoeuvres*"[3] according to Karin Sierralta, vice-president of the ISA and former president of the Peruvian surfing federation. This was one of the key arguments of the Castets project, which employed this rationalisation of training at a time when surfing was entering the Olympic Games. Project promoters envisioned the Castets pool as becoming a true national centre for high-level surfing, enabling surfers to train all year round when conditions in the natural environment are unfavourable. Similarly, the "Académie de la glisse", the name given to the Canéjan project, uses the same sporting argument. The performativity of the proponents' discourses, which describe "*artificial waves [as] an essential element for performance in the future*",[4] enables the promoters to establish the legitimacy—in this case sporting—of their projects.

A study of the social networks and arguments shared by the project promoters confirms the prominence of this element in their discourse. For example, this type of argument is legitimised by an executive from the French Surfing Federation (FFS) who was interviewed on this subject, for whom artificial waves allow "*the repetition of the gesture […], to be transferred to the natural environment*". Similarly, in his view, without this "*tool, we run the risk of finding ourselves in difficulty [compared to other nations]*". For him, the sporting usefulness of wave pools is not in doubt, and the risk lies in not embracing the technology which it is argued would precipitate France's competitive decline compared to other major surfing nations. In addition to high-level sport, the French Surfing Federation believes the Castets project would enable "*children and beginners to learn … in controlled conditions and in complete safety*",[5] at a time when surfing is becoming increasingly popular in France. Although it is difficult to obtain precise figures due to the high level of free (non-federal) surfing, the FFS predicts that there will be 680,000 surfers by 2020, with a 100% increase in the number of members in 10 years.

As well as sporting and economic arguments, the promoters of wave pools seek to justify their projects by enlisting the public support of well-known figures in the French surfing world. Their arguments are backed up by opinion influencers who have

[3] Interview conducted on 05/05/2023.

[4] Website: https://www.lemonde.fr/m-le-mag/article/2023/11/26/en-gironde-le-projet-de-surfpark-tente-de-prendre-la-vague-ecolo-pour-ne-pas-couler_6202429_4500055.html.

[5] Website: https://www.surfingfrance.com/actualites/federation/une-wavegarden-dans-les-landes.html.

the symbolic capital conferred on them by their social position, helping to establish their legitimacy in the media. Association leaders, surf champions and local elected representatives are all individuals who are approached by the promoters, who know how to mobilise their social capital, to legitimise and defend their interests in building wave pools. The Castets project was openly supported by the FFS, then chaired by Jean-Luc Arassus, for whom the existence of such a facility would have enabled "*the most experienced surfers to train [...]*".[6]

Norbert Ducrot, now retired from *Airbus Helicopters*, where he was head of commercial relations with Asia, is a case in point. The promoter and main funder of the Castets project, he regularly talks in the media, without putting forward any precise figures, about the "*jobs for young people*" that the sports facility would create. Ducrot also promotes it as an inclusive project, particularly in favour of disabled people, and justifies these claims by pointing to the interest shown in the facility by the Handi Surf association who sees it as a "*project [...] that will help to erase differences*".[7] The Canéjan project has the support of a number of French champions, including multiple world champion bodyboarder Pierre-Louis Costes, a deaf professional surfer and an amputee surfer, whose backing helps to justify the social utility of such a project from the point of view of "*accessibility*" and "*democratisation*", terms that are often used in their testimonials as part of the promoters' communication strategy on social networks. Championing these projects' inclusiveness is a novel messaging strategy compared with previous efforts, which focused solely on the sporting and economic aspects.

Lastly, although these projects are privately financed, they would have economic multiplier effects at the local scale and therefore attract the support of public authorities. For the mayor of Castets, the wave pool is seen as an economic catalyst for the area, transforming the commune "*from a town where you pass through to a town where you stop*".[8] For Norbert Ducrot, the pool's role would be to develop the local economy:

> So my project is a growth project. Economic growth, it's going to hire people, for every euro spent in the park you get two euros spent outside the park, that's it. It's going to attract 300, 400, 500,000 people a year. We're doing business. People are making a living from it. We're creating growth.[9]

According to him, wave pools would enable surf schools to work all year round and not be dependent on oceanic wave conditions, particularly in winter. In France, tourism jobs in coastal areas are structured by the length of the tourist season (Guibert, 2024). On the Atlantic coast in particular, service jobs in tourism—and therefore surf schools—are mainly characterised by fixed-term employment contracts during the summer period, rarely exceeding three to four months between May and September.

---

[6] *Idem.*

[7] *Idem.*

[8] *Idem.*

[9] Interview conducted with Norbert Ducrot, 15/05/2024.

However, the director of EuroSIMA believes that France's attractiveness and leadership in the surfing world could change in a few years' time if local authorities do not act in line with the global wave pool trend:

> But we also have to think about the attractiveness and national competitiveness if we are to maintain our leadership of the European surfing scene. And today we have to realise that we are in competition with Portugal and other countries that are going all out for international surfing competitions and surfcamp-type infrastructures, etcetera. [...] But we also have to bear in mind that in almost 5-10 years the cards may change and the situation may also change in Europe. [...]So if you like, we'd like to encourage local authorities to take action in this direction, which will also help to make the industry more competitive and attractive in our region.[10]

This observation is shared by Ducrot, who also sees these projects as genuine local resources: "*So today they prefer to take athletes, put them on a plane, go to Switzerland and train on the artificial wave in Switzerland. That's what the FFS does.*"[11]

However, these multiple benefits—sporting (high-level sport as well as sport for as many people as possible), social (jobs, sport for disabled people), economic (tourism spin-offs, opening up the region), etc.—are being called into question by the opponents, who are pushing back fiercely at local levels. One of the communication strategies used by the project promoters is to counter the opponents' arguments, in particular by mobilising the ecological argument: "*This project was designed and conceived with particular attention to environmental protection*",[12] said the promoter of the Canéjan project in October 2023 in response to the criticism levelled by local associations. However, the media and political resources of the opponents are so powerful in the local political arena that they have so far successfully contested the projects.

Environmental Criticism and Support from Associations and Politicians

One of the characteristics of wave pool protests in France is that they bring together heterogeneous individuals and social groups from different social and residential backgrounds, who have contrasting relationships with the disputed projects, and who are not involved with the same intensity, over the same length of time and sometimes for the same reasons (Dechezelles & Olive, 2017). They may contest because they were born in the territory or live there, because they consider the site to be ecologically or culturally exceptional or remarkable, or because they consider the wave pool projects to be ecologically and/or economically harmful. Opponents of wave pool

[10] Interview conducted with Franck Laporte-Fauret, EuroSIMA CEO, 10/05/2023.

[11] Interview conducted with Norbert Ducrot, 15/05/2024.

[12] *Sud-Ouest*, « Surfpark à Canéjan: comment les promoteurs du projet répondent aux opposants sur les réseaux sociaux», 15/11/2023.

projects, whether they are local residents or members of local or environmental associations, participate in the construction of a "reactive territoriality" (Lolive, 1999) where the territory defended constitutes a common good.

While the entrepreneurs behind the wave pools promote greater accessibility to surfing and its supposed democratisation, the prices charged in existing structures abroad are restrictive. In Switzerland, a 55-min session costs €110. Criticising the social and ecological dimension, an opponent of the Canéjan project, who is also a member of the environmental association Surfrider Europe, asserts that in his view these facilities would be "*[reserved] for the few and yet at the expense of a waste of energy and resources that are today unacceptable and illegitimate in view of the physical situation of the planet*".[13] The Canéjan area would also be negatively impacted, according to the association "Non au surf en boîte" (No to canned surfing), which was set up to fight the wave pool project in Canéjan: "*Nuisance for local residents: car traffic (~ 300,000 visitors per year) without public transport, undersized car park, loudspeakers, noise and vibrations from engines, events (concerts, shows, competitions)*".[14]

The use of this type of discourse on the negative repercussions for the area makes it possible to mobilise part of the local population by warning them of the nuisances caused by tourism (Christin, 2017). The resident populations have in fact taken psychological ownership of the public assets (lake, park) within the communes that are likely to be affected by these developments (Peck et al., 2021). From this perspective, the possible environmental impact of the Canéjan wave pool on the nearby watercourse is regularly raised by associations as a way of making local residents understand how the facility could affect their natural living environment: "*The Eau Bourde river would receive wastewater drainage and overflows from the site without any ecotoxicological studies having been carried out to assess the consequences of these discharges on the river and its ecosystem*".[15] These statements about the environment are legitimised by the professional roles held by the members of the collective: the spokesperson is also an ecological engineer at INRAE (France's National Research Institute for Agriculture, Food and Environment). These project opponents are therefore leveraging their status and skills in this local political field.

Following the same processes as the promoters, the opponents are mobilising opinion influencers to take the fight to the next level. "*Leisure projects don't get much press in France. There is a very, very vocal minority in France who will prevent this type of project*".[16] To this end, the citizens' mobilisation in Canéjan reached the local

[13] Interview with an opponent of the Canéjan project, member of the Surfrider association.

[14] Website: https://www.nonausurfenboite.fr/post/communique_reu_publique_28/. This association lists all the press releases and actions taken against the project on its website. Leaflets, posters and logos are also available.

[15] Website: https://www.cade-environnement.org/2023/06/10/piscines-a-vagues-de-canejan-le-cade-appelle-a-signer-la-petition/.

[16] Interview with Thomas Denis, promoter of the now abandoned wave pool project in Nantes, 02/05/2023.

member of Parliament, who took the fight all the way to the Senate.[17] The media coverage of the struggle in the regional and local press, which has already been used for other contested tourism projects (Clarimont & Vlès, 2016; Oiry, 2015), was also used in the context of this struggle, even reaching the national press, where the French Minister for Ecology openly questioned the justification for the wave pool project in Canéjan.[18] The positions taken by a senator and the minister, even if they are not definitive, ensure that the project is widely publicised in the extra-local media.

The spontaneous representations about surfing (Guibert, 2020), even if erroneous or exaggerated, are still prevalent, as shown by the speeches of opponents to wave pool projects who invoke the "roots" of the supposed culture of surfing and its relationship with nature. For one opponent, a member of an association fighting against the Canéjan project, the argument is based on the quasi-mystical relationship that surfing has with Polynesian roots and nature:

> Surfing is not a sport, it's not an individual ego competition. It's a spiritual prayer to the ocean in the great Polynesian tradition. Surfing the wave means respecting the ocean, thanking it for welcoming me [...]. It's a direct connection to the element on which Polynesian communities depend first and foremost.
>
> From the moment that sport became an Olympic practice, but in a natural environment, this type of project supported by the Ministry and the State stopped overnight. This has made local authorities, regions, departments and municipalities realise that today surfing should only be practised in a natural environment and that it is an aberration to do so in an artificial environment.

The reminder of the Polynesian origins of surfing, as well as the description of competitions as "*nonsense in terms of the original culture of surfing*" in the same interview, are important elements of the opponents' discursive strategies. In their view, these wave pool projects are denaturing the sport. According to an opponent from another association, they are also "*an anti-nature project*". By referring the mythical origins of surfing and its supposedly privileged relationship with nature, opponents were able to establish a normative and angelic discourse that placed surfing exclusively in a natural environment and far removed from a technologically controlled competitive practice.

Conclusion

Ecological harm and poor management of natural resources, an affront to the practice and culture of surfing, and the reproduction of social inequalities are arguments put forward in opposition to wave pools. These contrast with those in favour

---

[17] Website: https://www.senat.fr/questions/base/2023/qSEQ230607291.html#:~:text=%C3%80%20Can%C3%A9jan%20ce%20projet%20serait,loin%20du%20z%C3%A9ro%20artificialisation%20nette.

[18] Website: https://www.sudouest.fr/gironde/canejan/projet-de-surfpark-en-gironde-le-ministre-de-la-transition-ecologique-reconnait-les-interrogations-legitimes-des-opposants-16137067.php.

of the projects who speak of relieving congestion in natural surfing spots, making surfing more accessible and rationalising the training of top-level athletes. In the final analysis, when it comes to wave pool projects in France, surfing is used to oppose quasi-binary worldviews. Supporters and opponents are fighting over the legitimate definition of what surfing should be and, more generally, the future of surfing.

French opponents and proponents structure political positions regarding the development and construction of wave pool facilities. The ever-increasing importance of ecological issues and nature as a "common good" in the media, and therefore progressively in the representations of local populations, is making it difficult to build these facilities. In July 2023, associations opposed to the Canéjan surfpark project lodged an appeal with the Bordeaux Administrative Court, which suspended the project. In January 2024, further evidence relating to alleged health hazards associated with water quality and management was submitted to the court in an attempt to have the building permit cancelled, despite the fact that it had initially been signed by the local authorities.

As a result, political choices and positions have changed, and projects have even been abandoned altogether. In April 2023, following the abandonment of the Castets project, Ducrot signed a contract between his company "*Wavelandes*" and "*Power China Guiyang*" to build four surfparks in China. So, while these waves could be used as territorial resources (Guibert, 2006), citizen mobilisation is having an impact on the political support given to these projects. The change in the FFS's official position is revealing of this shift. Having openly supported certain projects, it gradually withdrew its support following an announcement that the Canéjan project's water consumption was higher than initially indicated by the project promoter. The challenge for the Federation is to avoid getting involved in a strategy that is too divisive within the surfing communities, even if one federal executive says he is personally in favour of it "*at the risk of attracting the wrath of certain environmental associations*". In fact, the project promoters feel that in addition to the pressure exerted by environmental groups, the problem lies in the lack of support from the political authorities who may not have the will to implement these projects, either because of the pressure exerted by opposing groups in the local political arena, or because certain regulations do not allow it. The promoter of another aborted project, near Nantes in north-west France, explains that although his project was initially fully supported by the town council, the municipal elections were the political moment that led to the council about face. Hoping to garner support of the local ecologist parties for reelection, the municipal council team reversed their support for the project perceived as environmentally harmful and therefore politically risky. The words of the Castets developer also reveal the importance of institutional support: "*First of all, institutional investors, you can't do anything without an institutional investor. People criticise the Chinese and other systems... But that's how things are in France, i.e. without the Banque Publique d'Investissement, without the Caisse des Dépôts, you can't do anything. Because the private investor asks" but is the State supporting you*?» In France only one project currently seems to be succeeding in overcoming the obstacles thanks to considerable support from local authorities and public institutions at regional and national levels. The project, "Okahina Wave",[19] near Poitiers, is due to open in 2025 on an existing lake. According to the promoter, "*semi-natural*" waves will form around an artificial atoll. One of the main arguments is that this artificial wave is environmentally friendly: "*The natural functioning of atolls has given birth to the first environmentally friendly surfing wave*". In this way, discourse around Okahina Wave attests to the project's ability to reconcile economic interests with respect for environmental constraints, since it is a "*new generation of artificial wave that reconciles economic development with respect for, and even development of, biodiversity*". If this project comes to fruition and proves to be a good use of resources, it's likely that other wave pools will be built by other stakeholders in France.

Although the debates and political controversies are sometimes heated at the local level between the opponents and supporters of these projects, the fact remains that, ultimately, wave pools are subject to democratic processes in which the mobilisation of citizens and associations thwarts the classic functioning of local political fields. In the past, even though social movements have always existed since the industrial era, decisions for the territories were mainly discussed between political professionals. The relationship between local residents to politics in France, without taking this as an essentialist vision, is undoubtedly marked by a "potential for conflict" within the territories in which the wave pool projects are a sort of manifest and proven trigger.

References

Audinet L., Guibert C., & Sebileau A. (2017). *Outdoor sports. A category of political action in question* (p. 108). Edition du Croquant.

Bourdieu, P. (1981). Political representation: Elements for a theory of the political field. *Actes De La Recherche En Sciences Sociales, 36–37*, 3–24.

Bourdieu P. (2000). *Remarks on the political field* (p. 112). Presses universitaires de Lyon.

Christin R. (2017). *Anti-tourism manual.* Écosociété, https://www.entrepotnumerique.com/p/59374?f=epub

Clarimont, S., & Vlès, V. (2016). Les contestations sociales du développement touristique dans les Hautes-Pyrénées: le rendez-vous manqué de l'innovation territoriale?. *Revue de géographie alpine*, 104-1.

Dechezelles, S., & Olive, M. (2017). Occupation movements: Acting, protesting, criticizing. *Politix, 117*, 7–34.

Dehez, J. (2020). The new resources of the surfing economy. In C. Guibert (Ed.) *The worlds of surfing. Historical transformations, social trajectories, technological bifurcations* (pp. 239–255). Editions MSHA.

Guibert, C. (2006). Communication policies and differentiated territorial identifications: The political uses of waves and the world of surfing by municipalities on the Aquitaine coast. *Téoros, 25–2*, 62–71.

Guibert, C. (2007). Le premier âge du surf en France: un sport socialement sélectif. *Science et motricité, 67*, 89–100.

[19] Website: https://www.okahinawave.com/fr/accueil/

Guibert, C., (dir.). (2020). *The worlds of surfing. Historical transformations, social trajectories, technological bifurcations* (p. 282). Editions MSHA82p.

Guibert, C. (2024). The meaning of work. Being a seasonal worker in a seaside resort in the summer. *Téoros* [Online], http://journals.openedition.org/teoros/12273

Lolive, J. (1999). *Protests against the Mediterranean high-speed train.* L'Harmattan.

Matonti, F., & Poupeau, F. (2004). Activist capital. An attempt at a definition. *Actes de la recherche en sciences sociales, 155*(5), 11.

Oiry, A. (2015). Conflits et stratégies d'acceptabilité sociale autour des énergies marines renouvelables sur le littoral français. *Vertigo, 15*, 3.

Peck, J., Kirk, C. P., Luangrath, A. W., & Shu, S. B. (2021). Caring for the commons: Using psychological ownership to enhance stewardship behavior for public goods. *Journal of Marketing, 85*(2), 33–49.

Saurugger, S. (2019). Interest group. In L. Boussaguet (Ed.) *Dictionary of public policy*, (5th edn). Presses de Sciences Po.

Subra, P. (2016). Chapter 3. Analysis of development conflicts: issues, actors, modes of action, representations. In P. Subra (Ed.) *Local geopolitics. Territories, actors, conflicts* (pp. 63–84). Armand Colin.

Tournadre, J. (2024). What is done here is gentler than in the ZAD. Politics of refusal in a contemporary communalist experiment. *Sociologie, 15*, 65–81.

Chapter 3
Integrating Science into Surf Wetsuit Design

Tyler M. Wiles, Jeff A. Nessler, and Sean C. Newcomer

Introduction

Nothing beats summer surfing where the only two required items are swimwear and a surfboard. Such was the life of a pre-1950s surfer who had the courage to surf in colder water without the use of a wetsuit. Until 1952, surfing was simple, the waves were uncrowded, and the most seasoned surfers could enjoy the sport deeper into winter if they were fortunate to live in a comfortable climate. A major milestone in surfing was the invention of the wetsuit that transformed surfing from a summertime activity to a yearlong, global sport. Despite the outsized contribution of the wetsuit to what surfing has become, very little attention has been dedicated to its technological evolution, nor the direction where future suits are heading (or could go). This chapter will explore the history, current technological trajectories, and potential implications of efforts to improve the efficiency, environmental friendliness, functionality, and effectiveness of the suits that keep us wet and warm while facilitating surfing longer sessions in more places worldwide.

The Warmth of Progress

The history of the wetsuit is rather ambiguous, unarchived, unverified, and travels by word of mouth rather than traditional record-keeping, possibly as a symptom of the relaxed culture that has persisted since surfing's inception. However, many online

T. M. Wiles
Department of Biomechanics, University of Nebraska at Omaha, Omaha, USA
e-mail: tylerwiles@unomaha.edu

J. A. Nessler · S. C. Newcomer (✉)
Department of Kinesiology, California State University San Marcos, San Marcos, USA
e-mail: snewcomer@csusm.edu

L. Mach (ed.), *The Future of Surfing in the Anthropocene*,
https://doi.org/10.1007/978-3-032-09377-6_3

articles, physical media, and discussions with long-standing industry professionals commonly point toward Hugh Bradner and Jack O'Neill as the fathers of the wetsuit in 1951–52. In collaboration with the United States Marine Corps, Bradner's goal was to design a replacement for early drysuits, believing that divers did not have to be dry to stay warm (the opposite of a drysuit) (Rainey, 1998). Early wetsuits were likely made from a unicellular foamed plastic material such as neoprene (invented by the DuPont chemicals company in the 1930s), rubber, or polyvinylchloride. Early wetsuit designs were also crude and inflexible. Hindered by slow progress, the original designs were quickly declassified for public aid and use but Bradner curiously decided not to pursue commercialization, citing that he presumed there were not enough divers and surfers in the world (Rainey, 1998). Lacking foresight on the wetsuit's potential impact on aquatic sports and the future growth of the recreational surfing population, Bradner left an opportunity that Jack O'Neill seized. Widely considered the entrepreneur who commercialized wetsuits specifically for surfing under his own brand, O'Neill drove the industry's development through his namesake company. Between the 1950s and the late 1980s, however, the surf wetsuit industry remained a slow-evolving niche market.

Up until the 1980s–1990s, neoprene panels were glued together forming rigid designs that allowed water to penetrate through the seams from gaps between the neoprene panels. Seam taping was introduced and alleviated some leaking issues, but the seams had to be protected and more durable. Stitching the panels together became common practice in addition to the advent of blind stitching. This method employs a curved needle that loops just beneath the surface of the neoprene without fully penetrating the material. This technique prevents the creation of holes in the neoprene, thereby stopping water from infiltrating the wetsuit. Most modern-day wetsuit designs still use these techniques and you'll often find that high end wetsuits are triple glued, blind stitched, and taped.

The last major change to neoprene itself was the inclusion of a Japanese super stretch formula in the 1990s. Neoprene was no longer rigid and inflexible, but incredibly stretchy and malleable to better fit the surfer's form. The ability to stretch the material was revolutionary and permitted easier access into and out of the suit. A new stretchable material also made thicker wetsuits feasible, and modern wetsuits can range from 2 mm all the way up to 8 mm for use in temperatures below freezing. Today, neoprene remains the material of choice as an insulator for wetsuits because it is highly durable and maintains flexibility over a wide range of temperatures.

Unfortunately, neoprene, formally named polychloroprene, is a carcinogenic synthetic rubber that is not environmentally friendly or health conscious (United States Environmental Protection Agency, 2010a, 2010b). For surfers, wearing a wetsuit has no known detrimental health effects because its carcinogens become inert during the manufacturing process. However, workers involved in manufacturing polychloroprene are exposed to these carcinogens daily. Equally important is the environmental impact of polychloroprene, since its manufacture generates waste that must be properly contained and disposed of, and its finished product neither degrades easily nor is recyclable. Given these facts, many wetsuit companies are making steps towards using neoprene-free materials, naturally sourced rubbers, or increasing the effectiveness of neoprene recycling and repair to limit unnecessary waste.

Stemming from the new super stretch formula, it wasn't until the late 80s or early 90s where the last evolution in the wetsuit began: the exploration of the wetsuit's entry. Many companies had long formalized back zip designs that required a zipper running from the lower back up to the neck due to neoprene's stiffness. As material flexibility improved, many wetsuit companies transitioned to using a chest zip that incorporated a zipper running horizontally between both collarbones. A flap of the wetsuit originating from the upper shoulders is pulled over the head, which is pushed through the wetsuit's neck hole, and then secured at the upper chest. One of the earliest chest zip designs is said to have had a zipper running along most of the surfer's wingspan. Both the back and front zip wetsuits have gone through minor modifications overtime, but the wetsuit market was revolutionized by the introduction of zipperless wetsuits from Japan. Zipperless designs quickly gained popularity and also mark a significant turning point in the surf wetsuit industry. The original zipperless wetsuit had two overlapping neoprene flaps that the surfer had to pull over their head to get on, with some struggle, but some still use this concept today. Those zipperless designs have since evolved, are easier to use, have colloquially been considered more comfortable and warmer than chest and back zips, and are becoming increasingly common in the modern-day lineup.

By the end of the 90s, neoprene was frequently sandwiched between a laminated nylon knit on the interior of the wetsuit for comfort and the exterior of the wetsuit for protection against abrasions and cuts. Imagine the chafing before an inner lining made of fleece was used! Thankfully, over the years, numerous companies have introduced wetsuits with inner layers claiming to be the warmest, most comfortable, least abrasive, most flexible, lightest, and most environmentally friendly. Despite those unsupported claims, until recently, most technological advancements were warmly welcomed. The outer, protective layer is also an important component of the wetsuit that is rubbery and durable. Some original designs used smooth neoprene panels, and similar designs have been making a comeback in recent years to improve warmth at a cost of durability. The outer layer of modern wetsuits is typically a combination of durable rubber around the torso with a scuba knit fabric around the extremities. This outer layer also gives the wetsuit company an opportunity to add their own flair by including logos, a wide range of colors to make their athletes identifiable in the water, and pockets for car key loops in the modern era. A few unorthodox outer layers have become synonymous with certain surfers, eras, and wetsuit brands or models such as Kolohe's white wetsuit, Hurley's double stripes, or Rip Curl's retro neon wetsuits.

Despite their slow implementation, these changes to wetsuit design over the past 70 years have undoubtedly led to significant improvements in wetsuit performance in the areas of durability, flexibility, and insulation. Evidence from a recent field-based survey study, in over 900 male and female recreational surfers reported that on average surfers rate their thermal comfort while wearing a wetsuit between "just comfortable" and "comfortable" (Warner et al., 2023). Unfortunately, there are no historical data on wetsuit comfort for comparison. Anecdotally, however, if you ask any salty old surfer how cold they were when surfing in the original wetsuits as a grom, they likely would respond by saying that they do not miss that aspect of the 'good old days!' It is important to note that these advancements in wetsuit design occurred relatively slowly and that this slow growth can likely be attributed to (1) the monopolization of wetsuit manufacturing and its reliance on polychloroprene closed cell foam as an insulating material and (2) the wetsuit industry's use of anecdotal evidence and testing to enhance product design. The most recent wetsuit developments have reportedly been driven by testing new ideas on a whim with little to no catalogue of their success made available to the public. The small number of companies competing for the best wetsuit designs without collaboration certainly curbed rapid development seen in other areas of the surf industry like swimwear, fin, or board design. One can speculate that the infusion of independent scientific inquiry is necessary to accelerate innovation in the stagnant field of wetsuit design and the introduction of neoprene alternatives.

From Laboratory to Lineup

Scientific inquiry into surfing is relatively novel and much of the early research focused on the physiological and biomechanical requirements of both competitive and recreational surfing. Many of those studies characterized the activity profiles of surfers and reported that surfers spend the majority of their time (~90%) in the water either paddling or resting (sitting or lying) and only a small fraction of their time (~5%) riding waves (Bravo et al., 2016; Farley et al., 2012; Minghelli et al., 2017; Secomb et al., 2015). This information provides insight on why designing wetsuits for surfing creates a unique set of challenges compared to designing wetsuits for swimming activities such as diving and triathlons. Specifically, the dynamic nature of surfing exposes different parts of the body to air or water depending on the activity. For example, the lower back of the surfer is exposed primarily to the air during prone paddling and resting (depending on board volume), but is submerged while resting in the seated position. The heterogeneous interaction of body parts with these two mediums during surfing provides a unique challenge to wetsuit design because the thermal conductivity of water is 25% greater than air, which translates to convective heat loss being five times greater in water compared to air. Based on this information, a series of three studies were designed to test the hypothesis that skin temperatures under a wetsuit during surfing would differ across the body. The first two studies were conducted in the field over a two-year period and measured skin temperatures under a 2-mm wetsuit at eight separate anatomical locations (Upper Chest, Upper Back, Upper Arm, Forearm, Lower Abdomen, Lower Back, Thigh and Calf) during surfing in recreational men ($n = 46$) and women ($n = 27$) (Corona et al., 2018; M. E. Warner et al., 2019). Results from these studies demonstrated that skin temperatures significantly differed across the measured anatomical sites, with an average 10 °C difference between the lowest skin temperatures recorded in the lower leg compared to the highest skin temperatures recorded in the upper back. In addition, these studies provided the first scientific evidence that differences between anatomical location skin temperatures were related to the amount of time these anatomical locations were exposed to water and/or air. Given this information, the authors concluded that wetsuits designed for surfing should distribute thicker neoprene in anatomical locations exposed to water and thinner neoprene in anatomical locations exposed to air. Therefore, the traditional neoprene distribution pattern that utilizes thicker neoprene in the torso and thinner neoprene in the limbs is suboptimal and outdated.

Data from these two studies on the impact of wetsuits on skin temperatures during surfing also suggested that female subjects had significantly colder skin temperatures than men at several anatomical locations (lower back and lower arm), which led the authors to speculate that differences in body fat distribution between sexes may play a role in thermoregulation during surfing (Warner et al., 2019). Historically the design of men's and women's wetsuits have been identical with only slight differences to account for anatomical differences between sexes. Therefore, these findings had practical applications supporting the need for wetsuits designed specifically for men and women. To confirm this finding, a third study was conducted to test the hypothesis that women's skin temperatures under a wetsuit differed significantly from men's skin temperatures (Skillern et al., 2021). This study had a similar experimental design to the previous studies, but utilized a simulated surfing protocol in a flume to control for the potential differences in environmental conditions and exercise intensities that likely influenced the previous findings. Using this well controlled experimental design, male ($n = 20$) and female ($n = 13$) surfers wearing 2-mm wetsuits alternated between paddling (speeds 1.2, 1.4, 1.6 m/s), duck diving, and resting in an Endless Pool Elite Flume with a water temperature set at 15.6 °C. Findings from this study demonstrated that core temperature was not significantly different between sexes even though there were significant differences between sexes at the upper arm, abdomen, and thigh skin temperatures. However, on average, it is important to note the skin temperature disparities among the three specified anatomical sites for men and women were relatively small, amounting to less than 1 °C. Given these small sex differences, the authors of this study concluded "that future wetsuit design may not need to consider thermoregulatory sex differences" (Skillern et al., 2021).

In an innovative approach for examining thermoregulatory sex differences during surfing, the researchers questioned a subset of male ($n = 10$) and female ($n = 5$) participants about their thermal sensations throughout the simulated surfing experiment in the flume, to assess if their perception aligned with their measured skin temperatures. Perceptual information about thermal comfort is a critical component in wetsuit design but had never been scientifically investigated before this study. Unfortunately, the sample size in this study was too small to provide a conclusive answer to the question of whether thermal perception is linked to skin temperatures under a wetsuit during surfing. To address this question, a recent field-based study administered a thermal perception survey to over 900 recreational surfers that were wearing full wetsuits as they exited the water across California (Warner et al., 2023). Information obtained from these surveys suggested that surfers felt the coldest, wettest, and least comfortable under their wetsuits in the chest, upper back, lower arms, and lower legs. Those findings suggest that the surfer's perception of where they were cold under their wetsuit did not fully align with previously reported skin temperature data. The authors of this article speculated that the discrepancies between skin temperatures and perception may be a result of water infiltrating the wetsuit through access points. Specifically, they speculated that water intermittently flowing through the wetsuit's wrist, ankle, and neck cuffs, and the front or back zippers of the wetsuit, likely contributed to the surfer's perception of being wet, cold, and uncomfortable. Those findings suggest that water infiltration into the wetsuit plays a significant role in surfers' thermal perception and that designers should focus their attention on creating novel design strategies to cuffs and zippers of wetsuits to limit water access. For instance, recent research has demonstrated that forearm skin temperature under a wetsuit can be significantly increased by simply adding a Velcro strap around the wetsuit at the location of the wrist (Kellogg et al., 2020). This data suggests that wetsuit designers should focus attention on creating novel design solutions to reduce water infiltration through points of entry (cuffs and zippers) to enhance skin temperature and thermal comfort in the future.

Recently there have also been several published studies focused on the use of novel materials to enhance thermoregulation in wetsuit design. One such study investigated if embossing the outer surface of the polychloroprene closed cell foam led to greater insulation properties than the standard nylon outer surface (Smith et al., 2020). Embossing is a technique used by the manufacturers to create a smooth shiny finish to the outside of a wetsuit by applying heat and pressure. Prior to these studies, wetsuit companies historically claimed that embossing the outer wetsuit surface led to enhanced warmth compared to wetsuits that used standard nylon outer surfaces. To test this hypothesis, recreational surfers ($n = 78$) were asked to participate in a series of studies that measured skin temperature at four anatomical locations (upper chest, upper back, abdomen, and lower back following previous findings) during surfing under a custom wetsuit with the torso composed of half embossed and half jersey materials. Results from this study provided the first empirical evidence that skin temperatures were significantly greater under the embossed side of the wetsuit in anatomical regions exposed to radiant heat from the sun and intermittent water exposure. Unfortunately, embossed polychloroprene closed cell foam lacks durability which subsequently limits its appeal. In the context of wetsuit manufacturing, the development of a resilient material exhibiting comparable radiant heat absorption and hydrophobic properties to embossed polychloroprene closed cell foam would represent a significant advancement.

Silicone-coated neoprene is another novel material that meets many of these described requirements and was recently introduced to the wetsuit industry as an alternative to embossing. Research testing in the field and laboratory suggests that its insulation capacity is like embossed polychloroprene closed cell foam (Paterson et al., 2023). The authors of this research concluded "that [wetsuit] manufacturers should consider the application of silicone coating for specific regions of their wetsuits to increase insulation capacity and durability" (Paterson et al., 2023). Future wetsuit designs should take advantage of the sun's warmth by continuing to optimize the absorption of radiant heat through novel materials.

The wetsuit industry has also incorporated novel materials into the inner linings of wetsuits in an effort to increase their insulation capacity. For example, Hurley introduced a mylar inner lining (Thermo Light) that they marketed with higher heat retention. However, unpublished research in fourteen recreational surfers demonstrated that no significant differences in average skin temperature existed between traditional nylon and mylar inner linings of wetsuits across anatomical sites (Newcomer, Unpublished data). Shortly after these findings were presented to Hurley, they discontinued using this inner lining in their wetsuits. More recently, Billabong introduced a Graphene fleece inner lining that they also marketed as warmer than traditional wetsuit fleeces. Expectedly, a recent study demonstrated that skin temperatures under the graphene fleece inner lining were no different from those under the traditional fleece inner linings in 48 surfers participating in recreational surfing (Denny et al., 2022). Furthermore, those surfers reported that they could not perceive any difference in skin temperature and comfort between the two inner wetsuit materials. These research studies provide excellent examples of how limited empirical scientific information is used by the surf industry for the marketing of their wetsuits.

Over the past decade, many wetsuit companies have become interested in developing materials that can replace polychloroprene foam, the middle layer of the neoprene wetsuit. Polychloroprene is undesirable as a wetsuit material because of its environmental and health impacts: as a rubber it is difficult to recycle, produces several tons of non-biodegradable waste each year, and is produced in a process that uses known carcinogens that leak into nearby wastewater (U.S. Environmental Protection Agency, 2010; United States Environmental Protection Agency, 2010b). Over the last few years, the wetsuit industry has introduced products such as Yulex, NaturalPrene, and thermoplastic elastomer (TPE) as potential alternatives to neoprene that exhibit similar material and structural properties, are bio-degradable or easily recycled, and are non-carcinogenic (Drobny, 2007; Grady et al., 2013; McKeen, 2014; Princi, 2017). To date, TPE is the only material to be scientifically tested in a wetsuit with a group of surfers. When a TPE wetsuit was compared to a neoprene wetsuit of similar thickness, TPE resulted in significantly lower skin temperatures at several anatomical regions that were regularly submerged (Wiles et al., 2022). However, surfers indicated that they did not perceive differences in temperature between the two wetsuits, and no differences were detected in any of the other variables analyzed. This suggests that TPE may be a suitable alternative to neoprene, perhaps with additional design modifications such as increased thickness or changes in cell size.

This study comparing a TPE and neoprene wetsuit provides a template for future testing of new materials in surf wetsuits. Alternatives to neoprene must provide adequate insulation and minimize heat loss, but it is also important to consider the impact that a wetsuit might have on other aspects of performance. For example, wearing a wetsuit has been shown to increase muscle activation in the shoulder and change paddling dynamics, presumably due to an increase in resistance to motion (Nessler et al., 2015). Those results suggest that a wetsuit with increased mass or greater material stiffness may lead to greater muscle activity and fatigue. Based on those results, deeper analyses of material properties will likely be an important factor in the design of future wetsuits. The mechanical properties of neoprene such as its stiffness and strength can be precisely measured on the bench-top using a tensile test to failure. A recent study reported the material strength, stiffness, and density of various types of neoprene and suggested that minor changes to wetsuit design can lead to substantial changes in material behavior (Paterson et al., 2023). In particular, a sample of embossed or smooth neoprene with only one layer of nylon jersey stretched more easily than a typical neoprene sample with nylon jersey on both sides. More recently, wetsuits made of different types of foam were shown to exhibit different amounts of stretch at the hip, knee, and shoulder during typical surfing movements like paddling and popping up (Yildirim, 2024). While these studies provide initial findings, further research will be needed to evaluate the connection between wetsuit material properties and surfer performance.

Wetsuits also compress the body. Research has shown that compression garments can enhance muscle power output, economy of movement, and endurance (De Glanville & Hamlin, 2012; Jakeman et al., 2010; McManus et al., 2023). Compression also appears to have a lesser-known impact on human motor control by improving proprioceptive joint position sense at the upper and lower extremities (Ghai et al., 2024). Surfers wearing a wetsuit have been shown to slightly alter the trajectory of the hand during the paddling motion (Nessler et al., 2015; Wiles et al., 2022). Contemporary analyses of human movement across repetitive paddling strokes suggest that a wetsuit may also lead to greater variability in movement patterns (Nessler et al., 2015). Those results may be linked to changes in sensory feedback provided by the wetsuit's mechanical compression. Optimizing the variability of motion during repetitive movements such as paddling may help to reduce the risk of over-use injuries despite them being less frequent than acute injuries (Farì et al., 2021; Remnant et al., 2020). Therefore, compression should also be a consideration in wetsuit design.

It is important to acknowledge that the wetsuit research described in the previous paragraphs was the result of a relationship developed between a large surf apparel company that was dedicated to wetsuit innovation and two academics with a passion for surfing from California State University San Marcos (CSUSM). This surf brand was unique in that they were affiliated with a large sports apparel company that had the resources required to use scientific inquiry in apparel research and innovation. This model, that had been used for decades by traditional sports apparel companies, was adopted by the surf brand in an attempt to revolutionize wetsuit design. With financial backing and a collaboration with academia, this surf apparel company had the freedom to experiment. In addition, their experiments were done at relatively low cost because many of them were integrated into preexisting Biomechanical and Physiological laboratory curriculum at CSUSM to help students to apply information learned in lecture. Unfortunately, the sale of this surf apparel company in 2019 prematurely ended this partnership. This has left a significant void in wetsuit research because all brands currently operate independently under their own funding, which in turn makes testing and scalability a significant challenge. Looking ahead, it's crucial for the wetsuit industry to find the necessary resources to continue integrating scientific testing in the development of new wetsuit designs.

Forecasting Wetsuit Design

The future of wetsuits is exciting and poised for significant advancement. Moving forward, it is important to continue integrating scientific testing in the refinement of new ideas. To date, much of the data-driven research described here has focused on improving insulation characteristics of wetsuits. Future research should incorporate a broader focus on aspects of wetsuit design that may impact the surfer's health and performance. For example, researchers might focus on designs that reduce both acute and chronic injuries associated with surfing. Specifically, surfers are increasingly suffering from acute lower extremity injuries (ankle, knee, hip) associated with landing aerials (Furness et al., 2021; Langenberg et al., 2021; McArthur et al., 2020). To reduce these types of injuries, additional support—such as a knee brace—could be integrated into wetsuit designs to provide the necessary stability for elite athletes facing increasing physical demands. An integrated knee brace has already been made for the 2024 men's surfing world champion and could be useful for recreational surfers recovering from injury or looking to prevent future injuries. Wetsuits carefully designed at the shoulders could also reduce overuse injuries that have been reported to occur as a result of suboptimal paddling mechanics (Farì et al., 2021; Remnant et al., 2020). Further, wetsuits could be designed to feature grooves that aid hydrodynamics or employ pennate fiber alignment to enhance paddling power. Attempts to enhance paddling efficiency through 'fin' like ridges on the arms have been incorporated into wetsuit design in the past, but were not scientifically tested and short lived.

Wetsuits could also be designed to be more buoyant to help keep the head/body higher above water when a surfer falls unconscious. There are a few custom wetsuits designed for big wave surfers that incorporate impact padding, passive, and active flotation devices for emergency use. However, inflatable vests and wetsuits have been found to alter paddling kinematics (Nessler et al., 2017). Due to a wetsuit's natural buoyancy, there is also the possibility that paddling speed might improve similarly to swimming; however, these improvements are likely dependent on a surfboard's volume and have yet to be tested in surfing (Parsons & Day, 1986; Quagliarotti et al., 2023). The combined effect of these and other incremental changes could amount to creating a unique wetsuit that meets the individual thermoregulatory, biomechanical, and safety needs of the surfer. Custom wetsuits that perfectly match an individual's anthropometrics are also now possible. A well-fitting, tighter wetsuit that allows less water infiltration will be warmer and potentially more durable. When combined with future advancements in wetsuit technology, this could significantly enhance performance.

Wetsuit innovation will also benefit from more radical, high-risk/high-reward ideas that incorporate novel materials and technologies. In the future, the wetsuit industry will need to take inspiration from other fields of textile research, development, and application to make revolutionary changes in wetsuit design. Smart materials, for example, are used to delay the rate of temperature change in athletes and wick away liquid. The introduction of an alternative material that can be optimally pre-heated before exposure to cold water, combined with a pre-surf routine that takes advantage of the body's natural heat production, could maintain a surfer's warmth for longer than normally expected (Cook et al., 2024). Single and multi-use heat packs are commonly applied to garments used in extremely cold climates. This technology could be adapted to an active heating system integrated into the wetsuit fabric, and can be adjusted through the use of waterproof technology housed within a surfer's watch. Similarly, naturally occurring heat derived from the upper body as a byproduct of energy production during paddling could potentially be harnessed to warm the legs using an innovative heat redistribution system.

Nature can also provide inspiration for radical changes in wetsuit design through biomimicry. Otters and seals are aquatic sea mammals that often share the lineup with surfers. Their ability to effectively thermoregulate in cold aquatic environments could provide novel solutions incorporated into wetsuit design. For example, incorporating synthetic fur similar to that of otters and beavers into wetsuits could enhance insulation by both creating a barrier to water and trapping insulating air bubbles within the fur (Nasto et al., 2016). Insulation through the addition of air or gases has been reported to be extremely effective, and the loading of cells within neoprene with noble gases like Argon, Krypton, and Xenon improve thermal protection of the wetsuit transiently (Moran et al., 2018). Another animal inspiration for wetsuit design could be derived from the hagfish, which can generate a protective layer of slime when threatened. Future wetsuits could be made to apply a biodegradable layer of slime, similar to that produced by hagfish, that would act as an additional layer of insulation. The creation of these biodegradable insulators could eventually result in a biodegradable spray-on insulation, similar to spray on shoes emerging in the market today. Imagine arriving at your local surf spot on a cold winter morning and, before your session, spraying on a perfectly fitting, biodegradable insulator with superior thermal characteristics and elastic properties compared to neoprene. At the end of your surf session, you could spray an enzyme onto this biodegradable insulating layer, breaking it down into a liquid safe to enter the ocean through the storm drain near your car. This would certainly be an improvement over having to put on a cold, damp wetsuit that smells like urine on a chilly winter morning.

References

Bravo, M. M., Cummins, K. M., Nessler, J. A., & Newcomer, S. C. (2016). Heart rate responses of high school students participating in surfing physical education. *Journal of Strength and Conditioning Research, 30*(6), 1721–1726. https://doi.org/10.1519/JSC.0000000000001263

Cook, C. J., Serpell, B. G., Hanna, L. J., Fox, A., & Fourie, P. J. (2024). Heat attainment and retention in surfers with and without a land-based warm-up and accompanying passive heat retention. *Sports, 12*(9), 241. https://doi.org/10.3390/sports12090241

Corona, L. J., Simmons, G. H., Nessler, J. A., & Newcomer, S. C. (2018). Characterisation of regional skin temperatures in recreational surfers wearing a 2-mm wetsuit. *Ergonomics*. https://doi.org/10.1080/00140139.2017.1387291

De Glanville, K. M., & Hamlin, M. J. (2012). Positive effect of lower body compression garments on subsequent 40-kM cycling time trial performance. *Journal of Strength and Conditioning Research, 26*(2), 480–486. https://doi.org/10.1519/JSC.0b013e318225ff61

Denny, A., Moore, B., Newcomer, S., & Nessler, J. (2022). Graphene-infused nylon fleece versus standard polyester fleece as a wetsuit liner: Comparison of skin temperatures during a recreational surf session. *Research Journal of Textile and Apparel*. https://doi.org/10.1108/RJTA-07-2022-0079

Drobny, J. G. (2007). Recycling of thermoplastic elastomers. In *Handbook of thermoplastic elastomers* (2nd edn., pp. 317–318). https://doi.org/10.1016/b978-081551549-4.50017-7

Farì, G., Notarnicola, A., Paolo, S. D., Covelli, I., & Moretti, B. (2021). Epidemiology of injuries in water board sports: Trauma versus overuse injury. *The Journal of Sports Medicine and Physical Fitness, 61*(5), 5.

Farley, O. R. L., Harris, N. K., & Kilding, A. E. (2012). Physiological demands of competitive surfing. *Journal of Strength and Conditioning Research, 26*(7), 1887–1896. https://doi.org/10.1519/JSC.0b013e3182392c4b

Furness, J., McArthur, K., Remnant, D., Jorgensen, D., Bacon, C. J., Moran, R. W., Hing, W., & Climstein, M. (2021). Traumatic surfing injuries in New Zealand: A descriptive epidemiology study. *PeerJ, 9*, Article e12334. https://doi.org/10.7717/peerj.12334

Ghai, S., Nilson, F., Gustavsson, J., & Ghai, I. (2024). Influence of compression garments on proprioception: A systematic review and meta-analysis. *Annals of the New York Academy of Sciences, 1536*(1), 60–81. https://doi.org/10.1111/nyas.15144

Grady, B., Cooper, S., & Robertson, C. (2013). Thermoplastic elastomers. In *The science and technology of rubber* (4th edn., pp. 591–652). https://doi.org/10.1016/B978-0-323-35824-8.00024-4

Jakeman, J. R., Byrne, C., & Eston, R. G. (2010). Lower limb compression garment improves recovery from exercise-induced muscle damage in young, active females. *European Journal of Applied Physiology, 109*(6), 1137–1144. https://doi.org/10.1007/s00421-010-1464-0

Kellogg, D., Wiles, T., Nessler, J. A., & Newcomer, S. C. (2020). Impact of velcro cuff closure on forearm skin temperature in surfers wearing a 2 mm and 3 mm wetsuit. *International Journal of Exercise Science, 13*(6), 9.

Langenberg, L. C., Vieira Lima, G., Heitkamp, S. E., Kemps, F. L. A. M., Jones, M. S., Moreira, M. A. de A. G., & Eygendaal, D. (2021). The surfer's shoulder: A systematic review of current literature and potential pathophysiological explanations of chronic shoulder complaints in wave surfers. *Sports Medicine—Open, 7*(1). https://doi.org/10.1186/s40798-020-00289-0

McArthur, K., Jorgensen, D., Climstein, M., & Furness, J. (2020). Epidemiology of acute injuries in surfing: Type, location, mechanism, severity, and incidence: A systematic review. *Sports, 8*(2), 25–25. https://doi.org/10.3390/sports8020025

McKeen, L. (2014). 8. Thermoplastic elastomers. In *The effect of temperature and other factors on plastics and elastomers* (3rd edn., pp. 371–398). https://doi.org/10.1016/B978-0-323-31016-1.00008-8

McManus, C., Cooper, C. E., & Sandercock, G. (2023). The influence of compression tights on running economy varies by relative intensity. *International Journal of Sports Science & Coaching, 18*(3), 793–800. https://doi.org/10.1177/17479541221097961

Minghelli, B., Graca, S., Paulino, S., & Sousa, I. (2017). Time-motion analysis of young competitive surfers: Southern Portugal. *Journal of Sports Medicine & Doping Studies, 07*(03), 3–7. https://doi.org/10.4172/2161-0673.1000194

Moran, J. L., Cottrill, A. L., Benck, J. D., Liu, P., Yuan, Z., Strano, M. S., & Buongiorno, J. (2018). Noble-gas-infused neoprene closed-cell foams achieving ultra-low thermal conductivity fabrics. *RSC Advances, 8*(38), 21389–21398. https://doi.org/10.1039/C8RA04037K

Nasto, A., Regli, M., Brun, P.-T., Alvarado, J., Clanet, C., & Hosoi, A. E. (2016). Air entrainment in hairy surfaces. *Physical Review Fluids, 1*(3), Article 033905. https://doi.org/10.1103/PhysRevFluids.1.033905

Nessler, J. A., Hastings, T., Greer, K., & Newcomer, S. C. (2017). Wearing an inflatable vest alters muscle activation and trunk angle while paddling a surfboard. *Journal of Applied Biomechanics*. https://doi.org/10.1123/jab.2016-0248

Nessler, J. A., Silvas, M., Carpenter, S., & Newcomer, S. C. (2015). Wearing a wetsuit alters upper extremity motion during simulated surfboard paddling. *PLoS ONE, 10*(11), Article e0142325. https://doi.org/10.1371/journal.pone.0142325

Parsons, L., & Day, S. J. (1986). Do wet suits affect swimming speed? *British Journal of Sports Medicine, 20*(3), 129–131. https://doi.org/10.1136/bjsm.20.3.129

Paterson, M., Moore, B., Newcomer, S. C., & Nessler, J. A. (2023). Insulation and material characteristics of smoothskin neoprene vs silicone-coated neoprene in surf wetsuits. *Sports Engineering, 26*(1), 23. https://doi.org/10.1007/s12283-023-00417-5

Princi, E. (2017). 5. Thermoplastic elastomers. In *Applied plastics engineering handbook* (2nd edn., pp. 91–107). Elsevier Inc. https://doi.org/10.1515/9783110640328-005

Quagliarotti, C., Cortesi, M., Coloretti, V., Fantozzi, S., Gatta, G., Bonifazi, M., Zamparo, P., & Piacentini, M. F. (2023). The Effects of a Wetsuit on Biomechanical, Physiological, and Perceptual Variables in Experienced Triathletes. *International Journal of Sports Physiology and Performance*, 1–9. https://doi.org/10.1123/ijspp.2022-0029

Rainey, C. (1998). *Wet Suit Pursuit: Hugh Bradner's Development of the First Wet Suit* (SIO Reference). UC San Diego: Scripps Institution of Oceanography. https://escholarship.org/uc/item/7353g3dj

Remnant, D., Moran, R. W., Furness, J., Climstein, M., Hing, W. A., & Bacon, C. J. (2020). Gradual-onset surfing-related injuries in New Zealand: A cross-sectional study. *Journal of Science and Medicine in Sport, 23*(11), 1049–1054. https://doi.org/10.1016/j.jsams.2020.05.010

Secomb, J. L., Sheppard, J. M., & Dascombe, B. J. (2015). Time-motion analysis of a 2-hour surfing training session. *International Journal of Sports Physiology and Performance, 10*(1), 17–22. https://doi.org/10.1123/ijspp.2014-0002

Skillern, N. P., Nessler, J. A., Schubert, M. M., Moore, B., & Newcomer, S. C. (2021). Thermoregulatory sex differences among surfers during a simulated surf session. *Sports Engineering, 24*(1), 16. https://doi.org/10.1007/s12283-021-00353-2

Smith, C., Saulino, M., Luong, K., Simmons, M., Nessler, J. A., & Newcomer, S. C. (2020). Effect of wetsuit outer surface material on thermoregulation during surfing. *Sports Engineering, 23*(1), 1–8. https://doi.org/10.1007/s12283-020-00329-8

United States Environmental Protection Agency. (2010a). *Chloroprene Chemcial Assessment Summary: CASRN: 126–99–8* (pp. 1–30).

United States Environmental Protection Agency. (2010b). *Toxicological Review of Chloroprene (CASRN 126–99–8)* (Issue 126). www.epa.gov/iris

U.S. Environmental Protection Agency. (2010). *Summary Overview of Health Effects Associated with Chloroprene: Health Issue Assessment* (No. EPA/600/8–85/011F (NTIS PB86197662)). U.S. Environmental Protection Agency. https://iris.epa.gov/document/&deid=29697

Warner, M. E., Nessler, J. A., & Newcomer, S. C. (2019). Skin Temperatures in Females Wearing a 2 mm Wetsuit during Surfing. *Journal of Strength and Conditioning Research*. https://doi.org/10.3390/sports7060145

Warner, M., Nessler, J., Filingeri, D., & Newcomer, S. (2023). The characterization of thermal perception in recreational surfers wearing wetsuits. *Applied Ergonomics, 113*, Article 104108.

Wiles, T., Simmons, M., Gomez, D., Schubert, M. M., Newcomer, S. C., & Nessler, J. A. (2022). Foamed neoprene versus thermoplastic elastomer as a wetsuit material: A comparison of skin temperature, biomechanical, and physiological variables. *Sports Engineering, 25*(1), 6. https://doi.org/10.1007/s12283-022-00370-9

Yildirim, N. (2024). *An analysis of Wetsuit Stretch in Surfers* (Cal State). California State University, San Marcos. http://hdl.handle.net/20.500.12680/v692tf24w

Chapter 4
Surf Cameras: Mediating Temporal Sovereignty in the Age of Digital Capitalism

Tyler Poisson

Background

Surf cameras are video streaming devices that capture ocean conditions at surf breaks around the world. They usually transmit real-time, wide-angle, muted footage via functional video-player interfaces, which feature minimal timestamp overlays. Typically, surf camera livestreams are broadcast 24/7 from fixed positions with static orientations and zooms. Weather resistant cameras and standard cables connecting to sources of power and the internet comprise the immediate infrastructural bases for these livestreams. This infrastructure is often installed and maintained on, and by owners of, private oceanfront properties (*Host a Surfline Camera*, n.d.). As a result, implementations and specifications vary from one location to the next. For instance, some cameras stream from the beach in 4K, while others stream 480p video at 24 frames per second from a distance. Under normal circumstances, surfers in frame are anonymized in virtue of their digital granularity. Although, it is possible to identify individuals in the water via surf cameras in cases when their equipment or habits are distinctive (Mann & Warren, 2023). In more or less detail, all surf cameras communicate the rhythm and approximate size of waves and line-ups (see Fig. 4.1). The ocean is the primary subject of every surf camera livestream, by default. Aesthetically, from the perspective of someone watching a surf camera on their computer screen, the ocean can be seen oscillating between naval blues and flicking with wave-crest whites in the foreground of the gentler blues and whites of the sky. Contributing to their minimalist liminality, surf camera interfaces lack the affordances that are typical of social media livestreams. That is to say, they do not, as yet, support comment sections or reactions that facilitate real-time interactions between users. Nevertheless, surf cameras are capable of evoking a range of emotions, from disappointment to anticipation, depending upon the picture of the ocean that they relay to the user.

T. Poisson (✉)
University of Massachusetts Amherst, Amherst, MA, USA
e-mail: tpoisson@umass.edu

L. Mach (ed.), *The Future of Surfing in the Anthropocene*,
https://doi.org/10.1007/978-3-032-09377-6_4

Fig. 4.1 Surf at Agate Beach State Recreation Site in Newport, Oregon, as taken sometime in December, 2024 by Connor (a participant in this study) from a Surfline.com surf camera livefeed

Brief History of Surf Cameras and Surfline

Sean Collins, who founded the for-profit surf forecasting business Surfline in 1985 as a pay-per-call wave forecasting service before it became the website Surfline.com in 1995, is credited with setting up the first surf camera for the 1996 US Open of Surfing in Huntington Beach, California (Mach, 2014; Pinto, n.d.; *Sean Collins: 1952–2011*, 2019). Reportedly, he used security camera hardware, in tandem with computer software called "Snap and Send," to relay one still frame of surf every five minutes to Surfline.com visitors (DeNuccio, 2024; Pinto, n.d.; Surfline Ramps Up Global Surf Cam Network, 2019). In 1997, Surfline.com supported six unique "surf cams" (see Fig. 4.2) and 1.5 million page views per month (Host a Video Cam, 1998; *Live Video Cams on the Net!*, 1997). By 2009, the company hosted 58 cameras worldwide. In 2014, some five years later, they maintained a network of at least 117 surf cameras. Five years thereafter, in 2019, Surfline was streaming live footage from roughly 600 cameras, a number that has since multiplied to 950 with the help of technological developments, such as Starlink, which provides internet access to remote areas of the globe (DeNuccio, 2024). Initially, Surfline.com expanded their network of free-to-use cameras by offering online advertising to local businesses that installed and sponsored Surfline-owned cameras on private coastal properties (*Live Video Cams on the Net!*, 1997). Presently, Surfline paywalls some of its cameras and generates revenue through subscriptions. Although a limited number of Surfline's cameras can be viewed for free, in order to access most of them, and any at all free of ads, at the time of this writing users must pay $119 a year for "Surfline Premium", or more for "Premium Plus", the latter of which grants access to "Smart Cams" that use "AI technologies… to detect, track, analyze and measure insights that you are trying to gather when watching the cams, such as wave height, wave frequency, ridden vs. unridden waves, and number of surfers." (Smart Cams & Privacy, 2024).

Surfline's cameras are integrated into their surf forecasts, which consist of wave height and period, wind speed and direction predictions, as well as information respecting weather, tide, and swell, for thousands of surf breaks around the world (Surfline Ramps Up Global Surf Cam Network, 2019). Surf cameras help surfers plan when and where to surf by providing them real-time footage of existing conditions. Surfline.com visitors can access limited forecasts for their favorite surf spots for free, but premium users enjoy advanced forecasting up to 16 days into the future. Behind the scenes, Surfline processes enormous quantities of data and takes into consideration such sciences as bathymetry and tropical meteorology. Reflecting a persistent effort to increase its forecasting accuracy and to expand its business, in which surf cameras have long played a role, Surfline has acquired oceanographic

Fig. 4.2 Screen grab of an archived version of Surfline.com from 1997 showing a map of then-existing surf cameras (*Live video cams on the net!*, 1997)

forecasting businesses such as buoyweather.com in 2003 and MagicSeaweed in 2017 (Pinto, n.d.). The company also markets its surf cameras as devices to be used to record oneself surfing. Surfline Premium members can rewind livestreams up to five days into the past. Moreover, they can trim, save, and share clips of themselves surfing, all from within the Surfline app. In fact, the "Surfline Sessions" smart watch app automates this process, providing users with ready-made clips of rides that take place within view of a Surfline Camera (*Capture Your Rides on Video with Surfline Sessions*, n.d.). Finally, Surfline recognizes that their cameras have "unexpected" use cases beyond helping surfers plan and record their surf, revealing that U.S. Coast Guard and lifeguard agencies have used them for public-safety purposes in as much as amateur storm watchers, UFO conspiracists, and nature enthusiasts have used them for their own entertainment and edification (Surfline Ramps Up Global Surf Cam Network, 2019). Surfline publishes and monetizes video compilations of surf camera footage on YouTube. Though they normally upload clips of surfers, one particularly well received video titled "WTF Moments Captured By Surfline Cams," which has at the time of this writing 590 thousand views, comprises footage of unusual animal and atmospheric activity captured by their surf cameras (Surfline, 2019).

With over 5 million users and the largest commercial network of real-time surf cameras and forecasts, Surfline is the most popular surf-focused media company in the world (*Surfline (@surfline) Instagram Photos and Videos*, n.d.). Even though Surfline enjoys an enormous market share and a monopoly on surf cameras at numerous surf breaks in the United States, a smattering of seaside towns and small businesses continue to host free-to-use surf cameras, undersigned by local sponsors in accordance with the original model pioneered by Surfline, on their websites in a bid to attract tourists. One such example is the Warm Winds Surf Shop in Narragansett, Rhode Island, one of the more popular surf towns on the East Coast, which offers free access to live footage of the waves at Naragansett Town Beach (*Warm Winds Surf Cam*, n.d.). Surfline, which maintains a separate camera at Naragansett Town Beach, broadcasts the only livestreams of two nearby surf spots, First and Second Beach. Even still, free surf cameras can be found on less trafficked corners of the web. For instance, surfcams.net lists about 60 free surf cameras hosted by an assortment of providers, though they aren't as reliable or technically capable as Surfline's cameras (*Free Surf Cams at Surfcams.Net*, n.d.). Not listed on surfcams.net are even harder to find and less typical cameras, such as "The Lanai Cam" in Seaside, Oregon, which can be remotely controlled by users willing to wait in a queue[1] (*The Lanai Cam*, n.d.). Owing to the quality and quantity of their cameras and forecasts, however, a plurality of U.S.-based surfers depend on Surfline Premium for their livestreaming and forecasting fancies.

[1] Ethan, a participant in this study, brought the The Lanai Cam to my attention.

Prior Research

At this point in time, very little social science research has been conducted on surf cameras, and to my knowledge no studies to date have investigated how surfers make use of them, despite the fact that a majority of surfers report using them (Mach & Ponting, 2018). In a study on surf tourism, Mach et al. (2018) surveyed 3101 surfers and found that 60% of respondents used surf cameras and forecasts more than once a day, while 22% used them at least once a week. What's more, 68% of participants agreed that surf cameras and forecasts influence if and where they surf on any given day. However, despite their reported utility, only 54% of surfers agreed that surf cameras and forecasts contributed to their enjoyment of surfing (Mach & Ponting, 2018). While Mach et al.'s data set is especially valuable because it is the only one of its kind, little of use to the present study on surf cameras can be extrapolated from it, since their survey items couple cameras in with forecasts, such that it is impossible to reach any definitive conclusions about surfers' use of one independent from the other.

Nevertheless, another data point of interest stemming from their survey indicates that 59% of respondents believe that forecasts and cameras contribute to crowding at their local surf break. This is in keeping with media coverage surrounding surf cameras, which tends to highlight local surfers' protests against, and sometimes even vandalism of, cameras in an effort to safeguard their preferred wave from overcrowding on the part of non-locals (DeNuccio, 2024; Macleod, 2024; Mann & Warren, 2023). Localism, which is ordinarily understood as a collection of hostile behaviors and strategies employed by surfers who frequent a particular surf break in an effort to deter newcomers and tourists from surfing said break, has historically been an integral facet of surf culture and one that news and entertainment media never tire of portraying (Barilotti, 2010; Daskalos, 2007; Mach, 2014; Scheibel, 1995; Walker, 2005). Although he doesn't consider the implications of surf cameras, Daskalos (2007) makes an important and relevant contribution to the localism literature. He interviews old school surfers in Southern California who turn to localism in response to an influx of newcomers spurred by gentrification. Daskalos observes that the old-timers' traditional social order has been displaced by a materialistic and individualistic ethic reflecting "the larger social context of postindustrial America's wealthier cities" wherefrom newcomers arrive. Surf cameras are today much more prevalent than at the time of Daskalos's writing. As already alluded to, local surfers have vandalized cameras in an effort to protect their social order and territory from people lacking in experience (Mann et al., 2024). In addition to contributing to overcrowding, Mach (2017, p. 47), quoted in Mann et al. (2024), points out that cameras and other technologies lead to "more efficient surfing (knowing when and where to go), meaning, not wasting time at the beach if the surf is not running, more waves in more places, having more and better images of surfing, to share with more people and faster over the internet." Therefore, surf cameras concretize the competitive and individualistic ethic that Daskalos attributes to newcomers. That is to say, surf cameras empower nonlocals and newcomers to go-when-it's-good and to capture footage of themselves, rending prior knowledge, communication, and camaraderie less essential to the practice of surfing. A newly released report prepared for the Surfrider Foundation entitled "Surfveillance: Swellnet's SurfCam at Winki Pop" explores the nexus of surf cameras, surveillance, and public policy through a case study at Bells Beach in Australia, where the surf forecasting company Swellnet operates a controversial camera in spite of protests on the part of local stakeholders (Mann & Warren, 2023). The study highlights "how current laws fail to address tensions between local community interests, individual rights, [and] corporate data extraction for profit" facilitated by surf cameras (Mann et al., 2024). These tensions are reflected in crowded surf breaks, privacy infringements, and the commercialization of a sacred and shared coastline.

Finally, econometrics studies, which treat surf breaks as common-pool resources, have uncovered positive correlations between fierce localism and surf break quality on the West Coast (Mixon, 2014; Mixon & Cebula, 2024). Relatedly, Leon Mach has explored the relationship between technological development and surf tourism, and has argued convincingly for stake-holder governance as a means to foster sustainability in the context of globalized surf culture and practice (Mach, 2014; Mach & Ponting, 2018; Mach et al., 2018). However, the foregoing research has up to now focused on trends in geographic space, with surf tourism and localism studies concerning, respectively, movements across boarders and strategies for enforcing them. Research in other domains, such as machine learning, undergird Surfline's surfer-tracking technologies which, the company assures surfers, do not use facial recognition software (Mann et al., 2020; Smart Cams & Privacy, 2024).

Statement of Problem

Evidently, there are more gaps in the literature on surf cameras than sedimentations. One especially glaring absence is that of research on surfers' direct experiences with surf cameras. It is one thing to *assume* that surfers use the technology as intended, to plan their surf and record clips of themselves, and thereafter to survey surfers about their surf camera habits on the basis of this assumption. It is quite another thing to start with the question of (1) how, exactly, *do* surfers use surf cameras? This question is the starting point of the present study. Through in-depth interviews with three surfers, this study aims to answer the foregoing and two further questions, namely: (2) how do surf cameras alter surfers' experience of surfing, and (3) how do surf cameras influence the sport's subculture of localism?

This study, as any other, has a particular significance and utility. Surfing is, for many who partake including myself, more than a fun form of exercise. It is a hobby that is at one in the same time meditative and exhilarating; a ritualistic practice of freedom capable of synchronizing bodies and minds with rhythms of nature. Under present conditions, technologies such as surf cameras contribute to commodification and thereby threaten to disembed surfers from their spiritual relationships to the ocean. However, as Leon Mach has contested, discourse surrounding the technization of surfing risks perpetuating a hopeless technological determinism—the idea that technology develops autonomously, altering society in the process—at the expense of a more sophisticated, and I would argue more thoroughgoing, understanding of the coevolution of technology and society, which stands, in contrast, to provide for beneficial developments in the sport of surfing (Mach, 2014, 2017). Following Mach in this line of reasoning, the present paper investigates surfers' relationships to one of the most widely used and understudied forms of surf media, the surf camera livestream, in an effort to better understand the evolving dynamics of surfing as sport and culture in the digital age. Accordingly, the framework developed herein has implications for policy makers and stakeholders seeking to remove, regulate, and/or introduce cameras at surf breaks around the world.

Theoretical Framework

In her book *Pressed for Time: The Acceleration of Life in Digital Capitalism*, Judy Wajcman explores how "technologies only come to life and have meaning as people adopt and use them." (2016, p. 3). Accordingly, she argues that technologies can contribute to our collective sense of the accelerating pace of life, or enhance "time-pressure", just as readily as they can help us gain control over our time. For Wajcman, the issue of time is not only one of speed but also one of distribution. Her concept of "temporal sovereignty" expresses the observation that agents in the technological era have power over their time, but to varying degrees depending upon their social, cultural, and economic contexts. As a result, the ability to control and shape one's time, that is to exercise temporal sovereignty, is not evenly distributed under digital capitalism as we know it. Although she does not explicitly define "digital capitalism," readers gather that it is characterized by the ubiquity and centrality of digital technologies, and thus by globalization and datafication. Surfing in the age of surf cameras is surfing in the age of digital capitalism. And since it is so, surf cameras facilitate changes not only in our experience of space but also in our experience of time. In what's to follow, I filter interview data through the prism of temporal sovereignty, making reference to adjacent theory along the way. This lens projects a threefold schema based on a conception of time involving distinctions between the past, present, and future. More specifically, it is found that surf cameras interface with an "immersive present", a "fragmented past", and a "budgeted future". Each of these analytic categories function as a fulcrum on which to balance our understanding of how surf cameras, as digital technologies and forms of media more specifically, simultaneously help surfers on the one hand and nuisance them on the other, as revealed by the interviews under discussion. When we surf we immerse ourselves in the present. Surf cameras, though, attenuate the immersiveness of any one surfer's session in virtue of empowering others to immerse themselves. The fragmented past refers to our memories of surfing, which surf cameras incentivize us to chop up into clips on the one hand, but enable us to save indefinitely on the other. The budgeted future portrays the increasing time-pressure we face in our lives, which surf cameras help us navigate at the cost of commodifying our leisure (Table 4.1).

The central thesis is this: surf cameras mediate temporal sovereignty so that the loss of control over time in any one case is counterbalanced by newfound temporal control in another. This suggests that Surfline is not only in the business of saving surfers time with its surf cam-enhanced forecasts. Neither, however, is Surfline universally disembedding surfers from their spiritual relationships to the ocean, as its critics contend. The tentative findings presented herein suggest that Surfline's surf cameras are *redistributing* time among surfers. Following Mach through Wajcman, our findings contravene deterministic narratives, demanding careful consideration of the costs and benefits of surf media and technology under digital capitalism.

All things considered, it cannot be forgotten, as Fine (2021) reminds, that:

> Riding waves or surfing is an Indigenous form of knowledge that predates European colonization. When Indigenous spaces are colonized, locations such as the surf zone, the wave riding practices themselves, and the ways in which we discuss and represent them become sites of tension. To surf is to reference both Indigenous agency and a history of colonial obfuscation as themes of cultural seizure are enmeshed in surfing narratives. (p. 218).

Mark Rifkin, in his book *Beyond Settler Time: Temporal Sovereignty and Indigenous Self Determination*, offers a different definition of temporal sovereignty than Wajcman. Rifkin (2017) develops of theory of "settler time", writing:

> U.S. settler colonialism produces its own temporal formation, with its own particular ways of apprehending time, and the state's policies, mappings, and imperatives generate the frame of reference… More than just affecting ideologies or discourses of time, that network of institutionalized authority over 'domestic' territory also powerfully shapes the possibilities for interaction, development, and regularity within it. Such imposition can be understood as the denial of Indigenous temporal sovereignty, in the sense that one vision or way of experiencing time is cast as the only temporal formation—as the baseline for the unfolding of time itself.

The struggle for control over one's time exists in parallel to the more fundamental, if less immediate, struggle over the very definition and structuring of time. Analyzing surf cameras through the lens of Rifkin's notion of "temporal sovereignty" implies that cameras impose a particular conception of time over physical spaces—the world's surf breaks—and activities—surfing. In so doing, cameras disrupt the temporal formations of those indigenous stakeholders for whom surf breaks and surfing are culturally, epistemologically, and/or spiritually integral. For this reason,

Table 4.1 Surf cameras and their effects on temporal sovereignty

	Immersive present	Fragmented past	Budgeted future
Surf cameras increase temporal sovereignty by…	Motiving users to surf	Enabling surfers to record and save footage	Empowering surfers to squeeze in sessions
Surf cameras diminish temporal sovereignty by…	Undermining investments on the part of locals	Transforming memories of sessions into discrete highlights	Commodifying leisure

it is imperative that future studies of surf cameras consider alternative conceptions of time in relation, especially, to the process of colonization. This essential task exceeds the scope of the present contribution.

Methods

For the purposes of this project, I conducted in-depth, one-on-one interviews, with three surfers in my extended network. The participants comprising this convenience sample were recruited informally over the phone, following referrals from acquaintances. A convenience sample was deemed appropriate due to the exploratory nature of this study. I conducted semi-structured interviews to foster organic and relaxed conversation between myself and each participant, a manner of interaction that is typical of surfers and thus desirable to replicate in the context of this study. I refrained from extracting information in as much as possible, attempting instead to excite, fascinate, and learn from each of the participants in the course of a mutually constructive dialogue. I prepared an interview guide with the goal of addressing my research questions (which, again, are (1) how do surfers use surf cameras?; (2) how do surf cameras alter surfers' experience of surfing; and (3) how do surf cameras influence the sport's subculture of localism?). Among other things, I asked participants to describe in detail the last time they watched a surf camera: why did they, how did they, where were they, etc. I also shared a link to a surf camera livestream and asked interviewees to describe what they saw and how they felt while watching it. In addition to relying on Zoom's meeting and transcription software, I took notes of my own in order to record contributions that stood out to me in real-time. Overall, the choice to conduct semi-structured interviews paid off, in that the richest responses I received came from the many follow up questions I asked in pursuit of emergent lines of inquiry.

Participants

The first person I interviewed, Connor, lives on the East Coast of the United States, in Rhode Island. He has been surfing in Rhode Island for around six years. The second interview I conducted was with Ethan, who grew up surfing in Hawaii and presently lives and surfs in Oregon. Finally, I interviewed a surfer named Will. Will has been surfing for six or seven years, roughly the same amount of time as Connor. He learned how to surf on the East Coast, on Cape Cod, but he has also surfed and lived on the West Coast, in Oregon. Will, Connor, and Ethan are middle-class white males in their mid-twenties, a sociodemographic profile that is overrepresented in the sport of surfing in the United States, but one to which I belong and that overlaps the most with the network from which I drew my sample. Importantly, Will, Connor, and Ethan all use Surfline surf cameras in some capacity, though only Connor and Ethan subscribe as Premium members.

The present study runs up against numerous and obvious limits, too many to identify exhaustively. The homogeneity of the sample marks a significant shortcoming, given the complicated history of surfing as it relates to colonialism, and the fact that today, the set of all surfers includes to varying degrees every gender identity, physical ability, age, class, race, ethnicity and geography. Future studies should engage a more diverse, and ideally representative, sample of surfers. It bears emphasizing that *most* perspectives, including those of women, those of old-timers, those of non-white surfers, those of Surfline representatives, and so on are excluded from the present study. This, of course, greatly diminishes, if not precludes, generalizability, for it remains unclear whether and how contributions from these key groups might have influenced the forthcoming analysis. This study is also limited geographically and culturally. Even though the present sample is bicoastal, it represents only one country, namely the United States. Surfers' experiences with surf cameras in global majority countries, as in more remote locations without reliable access to technologies, are invaluable data points to anyone who wants to understand what surf cameras are used for and how they affect subcultural change. For all of its faults, this study is advanced modestly as preliminary; the framework I develop herein may, or may not, prove to be a useful starting or reference point for future studies.

The Immersive Present

The 1973 surf documentary *Crystal Voyager* begins with a reflection on time. "You might be in there for only a few seconds in real time, but in your head it goes on for hours," narrates George Greenough, the film's subject. "It's an experience that's hard to describe, riding inside the eye of a big, grinding wave. Often you're riding so deep inside the tube, you don't make it out. You take a terrible wipeout. But what matters is when you're in there. It's a timewarp when you're inside the wave. Time enters a space, a zone of its own. The only reality is what's happening right then." This works as well as any other introduction to the immersiveness of surfing. The practice, it could be said, expands one's sense of the present.

In the effects-based tradition of media studies, researchers have proposed that media consumption *displaces* other, perhaps healthier, activities. Surf cameras have the opposite effect on the participants in this study. Apparently, cameras motivate users to surf. Although cameras can discourage surfers from heading to the break, such as when they relay crowded or poor conditions, none of the surfers I interviewed highlighted such instances. In fact, the participants in this study emphasized the cameras' magnetic pull. Connor explains that, when he sees even the slightest hint of waves on the camera, "I want to go get in the water." Will concurs, recognizing that it is their purpose to encourage people to get off of their screens and into the ocean. However, Will worries that cameras motivate people to surf for the wrong reasons. Uneasily, he sees Surfline promising to connect people with nature, fulfilling that promise, but also profiting off of its fulfillment. Corporations such as Surfline benefit monetarily and "wholeheartedly" from the expansion of the "surf industry," according to Will. Even still, watching real-time footage of good waves during our interview left Will "immediately wanting to be there." Indeed, Connor sometimes passes time by checking in on surf cameras throughout the day, even in cases when he is not planning on surfing, simply because they excite him. These feelings of anticipation are borne of the fact that once surfers get in the water, they enter a flow state that can be called the immersive present. Ethan, Connor, and Will surf for more than just exercise, but struggle to put words to their experiences in the ocean. Ethan likens surfing to flying, while Will refers to his experience in near religious terms. Surfing is an exercise of temporal sovereignty in at least two senses: one, it is something that people do with their leisure time, or that time over which they exercise the most control, and two, in virtue of the fact that surfers are free to forget time while surfing. When he's in the water, Ethan reports that his experience is so "immersive" as to exist outside of time. This is a common sentiment shared not only by Will and Connor but by surfers generally. In summary, surf cameras can deflect surfers from the ocean, depending upon the images they transmit to users. However, they appear to be more likely associated in the minds of their users with the power to attract surfers to the ocean. Insofar as surf cameras motivate users to surf, they encourage surfers to act on their temporal sovereignty in such a way as to maximize it. After all, for Wajcman, the *quality* of one's free time matters as much to the realization of temporal sovereignty as its quantity.

But there is another side to the immersive present. Many people who use surf cams, as Will admits people like himself, bring an ahistorical perspective to the wave and thus surf without taking from time important lessons that only it can teach. Speaking on his experiences surfing Hawaii and the mainland West Coast, Ethan defines locals as "just people who like understand the spot and have put a lot of time in at a spot." Put another way, locals earn their distinction and learn *from* time. For both Will and Ethan, the proliferation of surf cameras has changed what it means to "understand" a surf break, a prerequisite to any surf session, since wave quality and size vary drastically from one spot and day to the next. All three interviewees study surf camera livestreams, comparing what they see to what the forecast says in an effort to learn more about their surf spot of choice through digital media packaged by Surfline and transmitted to their screens. Unlike Connor who finds that surf cameras serve their purposes well enough, Will admits glumly that in fact there's so much he cannot learn from the cams. In order to understand a surf break, Will and Ethan contend, no doubt to the would-be agreement of Connor, that one needs to surf the break in person over the course of years so as to figure out during which seasons and winds the waves tend to be good, and "to know [even] what animals are there, what fish" in the words of Will. If locals are surfers who have invested a lot of time into understanding a particular surf break as Ethan suggests, then surf cameras undermine their temporal investment in that surfers new to the wave can nevertheless glean enough information from a camera to know, if superficially, whether or not and when to surf. It is in this sense that the surf camera redistributes temporal sovereignty. By democratizing access for those who are new to a wave, the surf camera facilitates immersive presence for wider swaths of enthusiasts. But by empowering these newcomers, Will and Ethan recognize that people who have, of their own choice, dedication, and perhaps privilege, devoted time to a surf break, face more crowded conditions and thus a loss of control over the quality of their surf, despite their temporal investments. After all, modern American surf ethic generally prohibits two people from riding the same wave, so in a crowded line-up locals are made to wait longer in between waves and, therefore, to surf less.

The Fragmented Past

One of the ways in which Connor and Ethan use Surfline's surf cameras is, as advertised, to capture clips of themselves in the water. Connor has a habit of looking at his watch and recording a mental timestamp after catching a good wave, so that he can later rewind the live camera and screen record himself surfing. Though Ethan was less forthright about this, it would appear that he has done the same thing.[2] This in turn promotes mediated social interaction: Connor not only shares clips of himself with friends nearby and far away, but also scours for clips *of* friends. The videos he captures double as training materials and digital mementos. Without cameras, Connor reflects, he'd have no other way to record third-person footage of himself in the water, or to watch his friends surf. Surf cameras enable ordinary surfers to create memories in the form of videos; they afford nonprofessional surfers, those without camera persons, the opportunity to access and share recordings. These records are used variously in social, pedagogical, and personal contexts. However, in the process of providing surfers unprecedented access to past moments, surf cameras can influence the salience of events as they happen in the water.

Ethan recounts one instance where he made a nice maneuver "and I was like, oh, I want to check that out and looked at my watch and like tried to remember like the time the whole, the rest of the time I was out there." The presence of the camera paired with Surfline's rewind feature, therefore, can facilitate a fragmentation of surfers' memories of the near past, where what is most memorable is the timestamp of a good wave. Even Will, who doesn't have a Surfline Premium subscription, doesn't hesitate to glance at his watch after catching a wave only to later text a friend who does, asking them to backtrack on the live cam and send him a recording of his ride. The presence of surf cameras, then, invites surfers to relinquish control over past time. In other words, surf cameras and their affordances shape the salience of events as, and especially after, they happen, influencing how surfers remember their experiences in the water. Making matters worse, the footage is always incommensurate to the reality. Ethan describes watching replays of himself surf as "humbling". Will offers a nuanced perspective:

[2] Connor and Will both report having been spatially influenced by the camera, such that they positioned themselves in the water so as to be in view of it.

> The actual experience of riding the wave is... You know, you could feel like you took the biggest drop of your of your life. And you could feel like it was a huge risk and then the camera could replay that video and depending on the angle the wave can just look completely different than what it actually felt like. So I think, you know, in terms of it warping people's sense of what the experience is like, I definitely think that that does happen. But I mean, I feel like surfing and anybody's experience in the water can be interpreted a million different ways, and it's not necessarily about how big or how risky the wave looks. It just kind of is if you're enjoying your time in the water.

Will's commentary reminds that surfers, not surf cameras, are in control: one doesn't have to replay the waves they catch, just because it is possible to do so. In fact, whereas Connor makes practical use of the clips he captures, Ethan remarks that he'd be better off without replays. Going a ways to explain this sentiment, Will reflects fondly on his "raw" memories of surfing breaks before learning that surf cameras existed. One gets the feeling that Will resents the camera for tempting him to fragment his surf sessions, since the desire to see one's self surf on camera is hard to resist. "The presence of the camera," Will suggests, "makes your moment in the water captureable in a way that I didn't experience before." He continues, with a melancholy expression, "I would rather just experience it in its raw form going forward." It would therefore appear that surf cameras can penetrate moments of immersion in the process of fragmenting surfers' memories.

The Budgeted Future

The need to budget one's future time was a recurring theme across all three interviews. Connor summarized his use of surf cameras by saying "it really comes back to time for me." Surf cameras broadcast real-time conditions, making it possible to check the accuracy of a surf forecast and make an informed decision about whether or not to surf in the moment. As a 9–5 worker, Connor is afforded a limited, though no less than average, degree of temporal sovereignty. Compounding the issue, he lives 20 min from the closest surf break by car, which more often than not is unsurfable due to meager wave heights. There was a sense of desperation in Connor's voice as he described his surf camera habits. Rather than risk driving 40 minutes round trip only to find that he's unable to surf on account of poor conditions, Connor will consult Surfline's cameras and base his decision off of them. This is especially necessary when swells bring big waves during the work week, when the only windows for surfing are in the early morning and late evening hours. For Will, who also works a 9–5 job but lives an hour from the coast, surf cams prove equally important in his calculations about how best to spend the time over which he has control. And even Ethan, living within minutes from a steady wave, purchased Surfline Premium in a bid to "save time." In this respect, when used as tools to plan surf sessions, surf cameras help workers regain and augment temporal sovereignty. In so doing, they also compress geography, opening up surf opportunities for non-locals which were previously reserved for locals only. This in fact plays out on a global scale as technological advancements in forecasting, in tandem with extant transportation technologies, incite surfers with the means to do so to travel around the world to catch the best swells at any given point in time (Mach, 2014).

These use cases remind of David Harvey's theory of "time–space compression." For Harvey (2000), transformations in the processes of production, distribution, and consumption associated with globalism have the effect of shrinking distances and speeding-up transactions in order to maximize capital accumulation. In keeping with this trend, surf cameras "make a wide range of simulated or vicarious experiences of what the world contains available to many people," opening up "the image of places and spaces… to production and ephemeral use." (Harvey, 2000, p. 84) Instead of checking in with a friend, taking a trip to the nearest break, or studying buoy data in order to gauge conditions, surfers can pay a subscription to instantaneously access livestream footage of numerous surf spots. What's more, cameras allow for media-multitasking, such that surfers can check the waves at work or while they tend to some other task, as all of the interviewees in this study have done. There is a sense in which surf cameras digitally reproduce surf sessions, too; through the medium of a screen they provide audiences with vicarious experiences of one or more of the many surveilled surf breaks at once. All things considered, surf cameras invite surfers to turn their leisure time into productive time. Put another way, Surfline opens surf breaks up to new forms of ephemeral consumption from which it profits.

Nevertheless, Connor appreciates surf cameras for their capacity to connect him to the ocean in reality, and sees his Surfline subscription as invaluable on this score. Without it, he would be forced to "gamble" with his time and drive 40 minutes roundtrip merely to check out the surf. With Surfline, he ends up surfing more. Ethan feels neutral about Surfline's cameras and is actively debating the utility of his Premium membership. From what I could gather, his ambivalence is due not only to his proximity to surf, but probably also to the fact that, as a West Coast surfer, he has seen firsthand how surf cameras can be received negatively by locals. Will, in closing, describes his discomfort respecting surf cameras. He criticizes Surfline for "commodifying quite sacred places" for indigenous communities around the world and provided the example of Bell's Beach, which is the subject of the aforementioned "Surfveillance" report. To quote Will, "the commodification of a beautiful site. I think in the long run, that's probably the most negative aspect of [surf cameras]. They drive people to locations basically with like short-sightedness about what the location really truly is." Surfline has a budgeted future of its own, which undoubtedly involves further expansion and commodification of its camera network and user base, in keeping with the demands of digital capitalism. Accordingly, it is to be expected that our uses of surf cameras will continue to alter the distribution of temporal sovereignty among surfers into the foreseeable future.

The Future of Surfing with Cameras

The astronomically ordained rhythms of the ocean place an upper bound on the period between swells, forcing the flow of capital into the sport of surfing to flood unsuspecting channels in fulfillment of its unrelenting mission to speed up time and infiltrate space. Surf tourism, with the help of advanced forecasting capabilities including camera networks, has globalized the sport (Mach, 2014). Furthermore, artificial wave pools make surfing possible where it wasn't before in physical space. Surf cameras facilitate changes in time, though we have seen that surfers' uses of cameras do not uniformly speed things up. Rather, they play a role in *redistributing* time along already existing poles. For example, they can be used to ease newcomers and intermediate surfers into the immersive presence of surfing, but in the process they are perceived to contribute to crowding, and thus diminish temporal investments on the part of old-timers.

The future of surfing is branching, and there is no telling which branch is ours. One possibility, pointed to by Leon Mach and others, is that surfing brings us together in an effort, or out of a need, to promote sustainable outcomes. Harvey recognizes that collectives of people "search for secure moorings" in response to time–space compression. He continues,

> place-identity, in this collage of superimposed spatial images that implode in upon us, becomes an important issue… Indeed, there are abundant signs that localism and nationalism have become stronger precisely because of the quest for the security that place always offers in the midst of all the shifting that flexible accumulation implies… [What's more] the assertion of any place-bound identity has to rest at some point on the motivational power of tradition. (Harvey, 2000, p. 89)

Surfers can turn towards the past, and to those old-timers who carry it with them in the present, in the process of bringing about a desirable future. Fine (2021) clarifies that "what we refer to as surfing, which is not used as a term until the twentieth century, is an encounter with the practice of *he 'e nalu* on the Hawaiian Islands. While it is its technological and commercial predecessor, it is also an attenuation of an ancient practice and form of knowledge" (p. 226). But how and what should surfers learn from those who carry knowledge traditionally associated with the practice of wave riding? What it means to adhere to surf tradition varies from one space–time point to the next. Furthermore, Harvey warns that "tradition is now often preserved by being commodified and marketed as such. The search for roots ends up at worst being produced and marketed as an image, as a simulacrum or pastiche." (p. 89) Historical practices and principles are not in themselves commensurate to the challenges of today, even when they are indispensable. For instance, traditionalism risks exclusion on the basis of gender, ability, race, and class. This does not preclude us from reorienting ourselves in community around the loci that sustain our habit of surfing in response to the disorienting forces of commodification and globalization. As they are presently used, Will points out that surf cameras *efface* knowledges of place which have been traditionally associated with surfing. But Connor's use of cameras reveals that they can also facilitate relationships *between* surfers. Moreover, there is no reason that cameras can't also be used as instruments of scientific analysis to *augment* our knowledge of coastlines and swells, particularly in the context of climate instability. In fact, Ethan in his interview describes triangulating camera footage, forecast data, and real time conditions in order to better understand the dynamics of his local surf break. Relatedly, Reineman (2016) argues that surfers are already positioned to contribute to costal research and resource management planning processes, thanks to their knowledge of surf breaks.

It cannot be forgotten that surf camera networks are predicated on human labor, from the workers who lay deep sea cable to those who manufacture camera parts and everyone in between. Presently, the chains of production and distribution that make surf cameras possible rely on materials and fuels whose extraction and use in turn harm the environment and the climate. It is inevitable that sea level rise, ocean acidification, and other interrelated crises, in tandem with our responses to them, will change what it means to surf (Sadrpour & Reineman, 2023). Data from surf cameras could prove useful to the task of modeling and protecting affected coastlines. But if opportunities to surf grow increasingly scarce under capitalism on account of such things as pollution and time-pressure, of what use will surf cameras be? Evers (2019) finds that for some surfers whose local breaks have been contaminated by plastic, radiation, and sewage, the practice of riding waves has already become a form of "polluted leisure" (see also David Brown's contribution to this volume). Should breaks like these, multiplying as they are, remain under the watchful eyes of surf cameras, surfers may turn to the camera not in an effort to judge whether the conditions are fair, but in an effort to discern whether it's *safe* to paddle out. In the absence of day-by-day and spot-by-spot pollution metrics, perhaps such data points as the color of the ocean will come to matter. In fact, Connor already uses surf cameras to spot Rhode Island's red tide algal blooms. Under such conditions, we can expect to see a proliferation of artificial surfing reefs and wave pools, unlocking new geographies and intervals for the surf market to exploit in the process of diverting increasingly scarce water resources towards increasingly expensive experiences (Guo, 2024). All else equal, we can also expect these manufactured landscapes to be surveilled by cameras. However, wherever surfing is stripped of its ecological context, the surf camera is likewise stripped of its functional disguise: conditions at artificial wave pools are consistent and amendable to our will, therefore cameras at artificial wave pools function not as forecasting tools but purely as surveillance devices designed to collect data so as to repackage it into commodities.

However, the future of surfing has less to do with the development of surf technology than it does with development at large. No one technology, and not even numerous in combination, can wholly determine our experience of the ocean, now or in the future. Instead, technologies accentuate already existing, ever in flux, interconnected social, cultural, political, and economic conditions, of which our experience is a culmination. For this reason it is hard to predict how surf cameras will coevolve with surfing. There are better and worse eventualities, to be sure, but wherever we as surfers end up on the dystopia-utopia continuum, it will not be *because* of surf cameras that we end up there. This much we've known, if not before, at least since Raymond Williams published *Television: Technology and Cultural Form* in 1974. Williams refutes technological determinism on the grounds that emergent technologies are themselves products of specific social, cultural, political, and economic contexts. For Williams, the television, and by extension new forms of media generally, establish new terrains with distinct topographies over which pre-existing social struggles evolve in novel ways. Therefore, there is nothing inevitable about how technologies, surf cameras included, adapt or are adopted. Indeed, their very existence is contingent on our activities. In accordance with Williams's thesis, and taking inspiration from Wajcman, the present study, preliminary though it is, reveals that surfers' uses of surf cameras redistributes time among and between them in accordance with preexisting subcultural, geographic, and economic conditions. Surf cameras are bound to continue to impact surfers far into the foreseeable future. Indeed, they already play a role in shaping that future. However, as Monique Mann, Leon Mach, and their collaborators demonstrate, communities are motivated to resist cameras when they are perceived to impinge on the rights of people and nature. As such, it is not clear whether surf camera networks are plateauing, whether they will expand to cover every conceivable break, or whether the near future will see fewer than exist today. Their effects on the culture and sport of surfing can be expected to sharpen and dampen in conversation with social and ecological trends owing to the process of capital accumulation, such as but not limited to wealth inequality and climate change. As Fredrick Jameson (1991) is so often paraphrased, it is easier to imagine the end of the civilization than the end of capitalism. Whether or not this is so, it must be easier to imagine the end of capitalism than it is to imagine the ocean waveless. If and under what conditions surfing takes place in a post-capitalist future is less up to our imagination than it is up to our collective commitments and practices, yesterday, today, and tomorrow.

References

Barilotti, S. (2010, July 22). *Localism Works*. https://www.surfer.com/news/plastic-polluter-1-coca-cola-is-not-your-friend

Beattie, K. (2001). Sick, filthy, and delirious: Surf film and video and the documentary mode. *Continuum, 15*(3), 333–348. https://doi.org/10.1080/10304310120086812

Capture your rides on video with Surfline Sessions. (n.d.). Retrieved December 12, 2024, from https://www.surfline.com/lp/sessions

Daskalos, C. T. (2007). Locals only! The impact of modernity on a local surfing context. *Sociological Perspectives, 50*(1), 155–173. https://doi.org/10.1525/sop.2007.50.1.155

DeNuccio, K. (2024, February 15). Did this website revolutionize surfing or kill a part of its renegade soul? Outside Online. https://www.outsideonline.com/outdoor-adventure/water-activities/surfline-forecasting-revolutionize-sport-of-surfing-or-kill-its-renegade-soul/

Evers, C. W. (2019). Polluted leisure. *Leisure Sciences, 41*(5), 423–440. https://doi.org/10.1080/01490400.2019.1627963

Elfick, D. (Director). (1973). *Crystal voyager* [Video recording]. https://www.surfertoday.com/surf-movies/crystal-voyager

Fine, H. H. (2021). Wave riding in cultural contexts: He'e Nalu, surfing, film, and discourse. *Pacific Asia Inquiry, 12*(1). https://www.uog.edu/_resources/files/schools-and-colleges/college-of-liberal-arts-and-social-sciences/pai/v12/11.Wave_Riding_.H.Fine.print_ready.10.10_copy_2.pdf

Free Surf Cams at surfcams.net. (n.d.). Retrieved December 12, 2024, from https://www.surfcams.net/

Guo, E. (2024). *The cost of building the perfect wave.* MIT Technology Review. https://www.technologyreview.com/2024/06/17/1093388/surf-pools-ocean-climate-change-water-scarcity/

Harvey, D. (2000). Time-space compression and the Postmodern Condition. In Held, D., & McGrew, A. (Eds.), *The Global Transformations Reader*. Polity Press.

Host a Surfline Camera. (n.d.). Retrieved December 9, 2024, from https://go.surfline.com/cam-host

Host a Video Cam. (1998). https://web.archive.org/web/19980109011943fw_/extremesportscast.com/cambiz.html

Jameson, F. (1991). *Postmodernism, or, the cultural logic of late capitalism.* Duke University Press. https://doi.org/10.2307/j.ctv12100qm

Jarman, D. (Director). (1993). *Blue* [Video recording]. Zeitgeist Films. https://zeitgeistfilms.com/film/blue

Live video cams on the net! (1997, January 25). https://web.archive.org/web/19970125041004/.surfline.com/videocam.html

Mach, L. (2014). *From the endless summer to the Surf Spring: Technology and governance in developing world Surf Tourism* [Dissertation].

Mach, L. (2017). Surfing in the technological era. In G. Borne & J. Gregor (Eds.), *Sustainable surfing* (pp. 41–71). Taylor & Francis.

Mach, L., & Ponting, J. (2018). Governmentality and surf tourism destination governance. *Journal of Sustainable Tourism, 26*(11), 1845–1862. https://doi.org/10.1080/09669582.2018.1513008

Mach, L., Ponting, J., Brown, J., & Savage, J. (2018). Riding waves of intra-seasonal demand in surf tourism: Analysing the nexus of seasonality and 21st century surf forecasting technology. *Annals of Leisure Research, 23*(2), 184–202. https://doi.org/10.1080/11745398.2018.1491801

Macleod, A. (2024). *(No) Empty set: The surf cam effect—Stab mag.* Retrieved December 10, 2024, from https://stabmag.com/premium/how-does-a-surf-cam-change-the-lineup-at-a-world-class-wave/

Mann, J. L., Angell, D. K., & Inan, N. C. (2020). *Final report: Identifying active surfers from surf camera footage to enable better surf forecasting tools.*

Mann, M., & Warren, I. (2023). *Surfveillance: Swellnet's SurfCam at Winki Pop.* Surfrider Foundation Surf Coast Branch.

Mann, M., Warren, I., & Baird, A. (2024). Surfveillance: SurfCams, social issues and regulatory intervention. *Journal of Sport and Social Issues, 48*(3–4), 224–244. https://doi.org/10.1177/01937235241276821

Mixon, F. G. (2014). Bad vibrations: New evidence on commons quality and localism at California's surf breaks. *International Review of Economics, 61*(4), 379–397. https://doi.org/10.1007/s12232-014-0205-9

Mixon, F. G., & Cebula, R. J. (2024). Seasons, sharks, and local control of the surfing commons: New evidence from the surf gangs of California. *American Behavioral Scientist*, 00027642241235832. https://doi.org/10.1177/00027642241235832

Pinto, L. M. (n.d.). *The story of Surfline.* Surfertoday. Retrieved February 7, 2025, from https://www.surfertoday.com/surfing/the-story-of-surfline

Reineman, D. R. (2016). The utility of surfers' wave knowledge for coastal management. *Marine Policy, 67*, 139–147. https://doi.org/10.1016/j.marpol.2016.01.023

Rifkin, M. (2017). *Beyond Settler time: Temporal sovereignty and indigenous self-determination.* Duke University Press. https://doi.org/10.2307/j.ctv11smrwm

Sadrpour, N., & Reineman, D. R. (2023). The impacts of climate change on surfing resources. *Shore & Beach, 91*(1), 32–48.

Scheibel, D. (1995). "Making waves" with burke: Surf Nazi culture and the rhetoric of localism. *Western Journal of Communication, 59*(4), 253–269. https://doi.org/10.1080/10570319509374522

Sean Collins: 1952–2011. (2019, December 18). Surfline. https://www.surfline.com/surf-news/sean-collins-1952-2011/73239

Smart Cams and Privacy. (2024, July 11). *Surfline Support Center*. https://support.surfline.com/hc/en-us/articles/27224444365211-Smart-Cams-and-Privacy

Surfline (Director). (2019, June 10). *WTF moments captured by surfline cams* [Video recording]. https://www.youtube.com/watch?v=8GwAqsbk7HI

Surfline Ramps Up Global Surf Cam Network. (2019, May 26). Surfline. https://www.surfline.com/surf-news/surfline-ramps-global-surf-cam-network/52678

Surfline (@surfline) Instagram photos and videos. (n.d.). Retrieved December 12, 2024, from https://www.instagram.com/surfline/

The Lanai Cam. (n.d.). Retrieved December 12, 2024, from http://68.185.24.173/local/popup4.html

Wajcman, J. (2016). *Pressed for Time: The acceleration of life in digital capitalism.* University of Chicago Press. https://press.uchicago.edu/ucp/books/book/chicago/P/bo19085612.html

Walker, I. H. (2005). Terrorism or native protest? *Pacific Historical Review, 74*(4), 575–602. https://doi.org/10.1525/phr.2005.74.4.575

Warm Winds Surf Cam. (n.d.). *Warm winds*. Retrieved December 13, 2024, from https://www.warmwinds.com/surf-cam

What are Smart Cams? (2024, July 9). *Surfline Support Center*. https://support.surfline.com/hc/en-us/articles/25643357257627-What-are-Smart-Cams

Williams, R. (1974). *Television: Technology and cultural form* (3rd edn.). Routledge. https://doi.org/10.4324/9780203426647

Part II
Environment

Chapter 5
Unconventional Waves on a Heating Planet

Scott Laderman

Tyler Wright faced an unusual predicament. In August 2024, the World Surf League (WSL) announced that Surf Abu Dhabi, a new wave pool in the capital city of the United Arab Emirates (UAE), would serve as a competition venue on the 2025 WSL Championship Tour.[1] For Wright, a two-time WSL world champion and the only openly gay surfer competing in those elite ranks, this presented a dilemma. Homosexuality and same-sex sexual activity were illegal in Abu Dhabi, and the penalty for violators could range from years in prison to possibly even death.[2] While there "have not been any documented prosecutions of consensual sexual acts" since 2021, Amnesty International reported in 2024, schools were warned in September 2022 to "refrain … from discussing gender identity, homosexuality, or any other behavior deemed unacceptable to UAE society," and "vague language criminalizing 'scandalous act(s) offending modesty'" remained on the books.[3] To be openly queer in Abu Dhabi thus carried risk. Wright had been outspoken in her political views and had worn the Pride flag on her jersey since 2020.[4] If she were to compete in Abu Dhabi, especially if accompanied by her wife, she could potentially face legal jeopardy.

The WSL's decision to host a competition in the UAE thus invited withering criticism from Wright's family, including her wife, Lilli Wright, and her brother Mikey, who is himself a former competitor on the WSL Championship Tour.[5] The WSL had "no business putting on an event at a location where my sister can be sentenced by law with the death penalty," Mikey Wright told the organization. "So much for equality and equal rights, only when it's convenient to WSL." Lilli Wright was just as scathing. "Even after winning 2 world titles she is still not valued enough by the WSL to be considered when they sold this event. WSL have the duty of care to their athletes to not put them in potentially life[-]threatening circumstances like this," she wrote on Instagram. "Tyler's queerness should not have to be a burden

S. Laderman (✉)
University of Minnesota, Duluth, USA
e-mail: laderman@d.umn.edu

L. Mach (ed.), *The Future of Surfing in the Anthropocene*,
https://doi.org/10.1007/978-3-032-09377-6_5

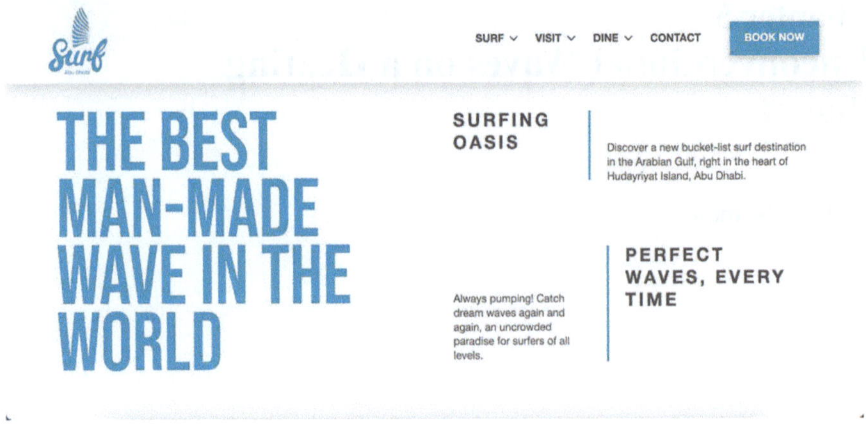

Fig. 5.1 Surf Abu Dhabi does not hold back, declaring itself home to "the best man-made wave in the world." It is "always pumping," its website boasts, with "dream waves" in "an uncrowded paradise for surfers of all levels." Website screenshot courtesy of Surf Abu Dhabi

or an obstacle in her workplace. … At the end of the day, WSL had absolutely no business selling this event to this location expecting their only openly queer athlete to go along quietly."[6]

That this whole episode even materialized is a result of a peculiar development: The WSL had scheduled a world-class surfing competition far from world-class ocean waves (Fig. 5.1).

It was able to credibly do so because of the remarkable progress that had been made in recent decades in wave pool technologies, as Jess Ponting details in Chap. 1 of this volume. From Surf Snowdonia, the Welsh pool that opened to the public in 2015, to Texas-based Waco Surf and Kelly Slater's Surf Ranch in California's Central Valley, wave pools had moved beyond the barely rideable slop of the twentieth century into hollow, picture-perfect walls in the twenty-first. To be sure, the WSL's earlier foray into wave-pool contests with the freshwater Surf Ranch Pro had been, according to one report, "possibly the most maligned event *ever* on the World Tour." But, that same report noted, the contest, which was first staged in 2018, was reportedly one of the few moneymakers for the World Surf League.[7] How could it therefore resist trying to cash in on the petro-derived wealth of the Emirati state?

The UAE apparently recognized the value in such a contest. In recent years a number of regional powers have used sport to try to burnish their global images. In late 2022, Qatar hosted the FIFA World Cup. Months earlier, Saudi-owned LIV Golf fielded its first event at England's Centurion Club. And in 2024, Riyadh welcomed the Women's Tennis Association Finals, with the Saudi kingdom offering the largest purse—some $15 million—in the history of women's tennis. "For years now," the *New York Times* reported, "Crown Prince Mohammed bin Salman, the country's de facto ruler, has used Saudi oil riches to remake his staid kingdom into a global player with glitz to match, pairing tectonic social changes with intensifying repression," including the macabre 2018 killing and dismemberment of *Washington Post* columnist Jamal Khashoggi and the political imprisonment of hundreds of Saudi nationals, including prominent women's rights activists.[8] Critics have regularly denounced the Saudi regime's "sportswashing."

Yet it's not just sport. Art and higher education have also played into these Gulf-state cultural-diplomacy campaigns. Several American universities opened branch campuses in Qatar (Carnegie Mellon, Georgetown, Texas A&M, Virginia Commonwealth, and Northwestern Universities and Weill Cornell Medicine), Abu Dhabi (New York University), and Dubai (Rochester Institute of Technology) despite sometimes strong faculty opposition, with free speech and academic freedom largely a mirage in what the American institutions hoped would be profit-generating outposts.[9] And in the arts, Abu Dhabi, in particular, has attempted to emerge as a significant global player, creating a world-renowned arts district on Saadiyat Island. The Louvre Abu Dhabi, which was designed by architect Jean Nouvel, was inaugurated in late 2017; by 2019, it had attracted two million visitors, becoming, it claimed, "the most visited museum in the Arab world."[10] Just down the road, the enormous, Frank Gehry-designed Guggenheim Abu Dhabi planned to open its doors in 2025.

In other words, whether it was sport, higher education, or the arts, the Gulf states saw cultural-diplomatic gold in presenting themselves as cosmopolitan contenders in the emerging world order. And for-profit entities, such as the World Surf League, showed no hesitation in trying to exploit the nascent opportunities these wealthy petrostates presented. For professional surfing, the emergence of machine-generated, flawless waves made this possible, enabling the WSL to host an exclusive contest in a wealthy Persian Gulf city better known for long-haul layovers than for quality surf. The fact that the Abu Dhabi pool used saltwater only added to its appeal, as it seemed to negate some of the worries that had been raised about environmental sustainability while addressing contestants' performance concerns.[11] (Saltwater has generally been avoided in wave pool designs because of the high costs associated with its corrosive effects.[12]) And lest one dismiss a surf contest as small potatoes in the face of soccer games in Qatar or tennis matches in Saudi Arabia, in 2017 the WSL "ranked third among sports leagues around the world in social interactions and video views, behind just the NFL and NBA," according to a study for *Sports Business Journal*.[13]

*****

Abu Dhabi may be among the most high-profile—and possibly the most lucrative—of these unusual surfing venues, but it is hardly the first. For decades surfers have in fact found waves in unconventional places. Wave pools may garner a disproportionate share of media attention, but they represent only a portion of the surfing done on non-oceanic waves. From the Great Lakes and the Snake River to Munich, the Severn Bore, and the Baltic, Black, and Caspian Seas, surfers have sought out waves wherever they can find them, producing vibrant surf cultures sometimes hundreds or even thousands of miles from the nearest ocean coast. Riding these non-oceanic breaks may sometimes require specialized equipment and greater patience—long waveless periods are not uncommon—but the rewards are tangible: Surfing has become accessible to people and communities that might not otherwise fit the traditional wave-riding profile.

There are sometimes political considerations that drive these developments—or at least political concerns have a tendency to impose themselves on them. This is the case with the aforementioned Abu Dhabi Pro, and the same might be said for the 2024 decision by China Tourism Group to partner with American Wave Machines in building a PerfectSwell wave pool and surf resort in Hainan. "With the launch of PerfectSwell, China solidifies its commitment to developing world-class athletes across all Olympic sports, surfing included," resort surf park director Kuan Zhao said of the nation's global sporting ambitions, which are deeply tied to issues of soft power and national prestige.[14]

But more often it's about fun. Surfers riding unconventional waves tend to do so for the same reason surfers on the ocean do. Whether it's the satisfaction that comes with overcoming one's fears, the thrill of testing—perhaps for the first time—an unusual wave, or the exhilaration to be found in harnessing the energy of a moving mass of water, surfing ultimately brings them pleasure. And there's clearly a lot of pleasure to be found. Professional surfer Dylan Graves predicated much of a popular three-season online series on scouting out and riding unconventional, or what he called "weird," freshwater or brackish waves.[15] Others have effectively done the same.[16] And in a nod to just how far surf culture had spread, in 2024 *The Inertia* published what it called "The Definitive List of Man-Made River Surfing Waves," which spanned ten U.S. states, one Canadian province, eight countries, two continents (Europe and North America, including Hawai 'i), and Oceania.[17] Readers identified yet others.

Given the changes wrought by the heating planet, what the future holds for some of these relatively recent surf zones is uncertain. While considerable attention has been focused on rising sea levels and what these might mean for coastal Florida, Southern California, or various Pacific islands, surfers far from the nearest ocean will not be immune from the effects of the climate crisis. How will seasonal river waves be affected if droughts result in reduced snowfall and thus to diminished spring melts? What will happen to the surf breaks on Lake Michigan or Lake Erie amid fluctuating water levels from torrential downpours or months or years without meaningful precipitation? How responsible or even feasible is it to build wave pools in arid regions in which fresh water is in short supply? These are just some of the questions that surfers far from the coast will have to consider.

Perhaps the oldest non-oceanic surf zone—and one that is already beginning to see the effects of the climate crisis—is the Great Lakes of North America. All five of these freshwater bodies are connected by a system of nineteenth- and twentieth-century locks, canals, and channels, and modern engineering has likewise connected the lakes to the Atlantic Ocean through the St. Lawrence River. All five of the Great Lakes have surf, though the quality of the waves and the months during which they are rideable vary. Lake Erie is the shallowest of the five, and Lake Ontario is the smallest by surface area. Lakes Huron and Michigan are technically one water body—they are connected by the three-and-a-half-mile-wide Straits of Mackinac—and Lake Michigan is probably the most frequently surfed, in considerable part because of the higher population density of its shoreside communities; these range from Chicago, Illinois, and Gary, Indiana, to Milwaukee and Green Bay, Wisconsin, and numerous beach towns on Michigan's Lower Peninsula. Lake Superior is the largest and deepest of the five lakes, and its water is by far the coldest.[18] In fact, it contains more water than the other four lakes combined; indeed, its volume is so great—approximately 10 percent of earth's surface fresh water—that Superior could fill all the other Great Lakes as well as three more Lake Eries.[19]

All five lakes rely on wind swell. They are not large enough for groundswell, so the lakes' fetch—the distance wind can travel across open water—is of critical importance. Lake Superior, with the coldest water and an east–west fetch of approximately 350 miles (and approximately 160 miles from north to south), tends to see the largest waves, though all five lakes are capable of producing big surf.

People have been surfing the Great Lakes longer than most oceanic surfers probably appreciate. The first known instance took place when Doc "Makaha Dave" Seibold, who had been stationed in Hawai'i during World War II, paddled his longboard into some Lake Michigan waves at Grand Haven in September 1955.[20] Just a decade later the Lake Shore Surf Club was founded in Sheboygan, Wisconsin, and the Great Lakes Surfing Association of Grand Haven followed a year after that.[21] Much of this early freshwater surfing took place in the relatively warmer climes of Lake Michigan, yet Lake Superior—the world's largest freshwater lake by surface area (and third largest by volume) and the one boasting the best wave quality in the interior United States—has received probably the greatest attention in recent years.[22]

There are small communities of surfers in the areas around Marquette, Michigan, and Thunder Bay, Ontario, but by far the greatest concentration of surfers—and the lake's premier break, Stoney Point—can be found in and around Duluth, Minnesota. The so-called Zenith City at the far western terminus of the lake, Duluth, together with neighboring Superior, Wisconsin, houses several colleges and universities, two major medical centers, and North America's farthest inland Atlantic seaport, which is the busiest on the Great Lakes. Among the city's 86,000 residents are dozens of surfers, who are regularly joined in the water by people from elsewhere in the Upper Midwest, especially the Twin Cities metropolitan area of Minneapolis and St. Paul approximately 150 miles to the south.

Lake Michigan surfers report having surfed Lake Superior in the 1960s, and Greg Isaacson of Duluth claims that he began surfing the lake in 1975.[23] Surfer Publications' *The Surf Report* even published a Great Lakes issue that included Minnesota—though in somewhat foreboding terms—in 1993.[24] But the wave riders who set Superior's flourishing contemporary surf culture in motion were undoubtedly Bob Tema and Brian Stabinger (Fig. 5.2).

Tema, who grew up in Hawai'i before relocating to Minnesota in the mid 1990s, began exploring the lake with Stabinger in 1998.[25] They quickly came to appreciate that Superior is strictly a cold-water zone. Summers are almost entirely flat, with the surf season typically beginning in the autumn and extending through the spring. Given the region's brutally cold winters, in which water temperatures drop into the low 30s and air temperatures periodically hover below zero, this means the thickest of wetsuits, booties, mittens, and hoods. I first surfed Lake Superior when I accompanied

Fig. 5.2 Bob Tema, together with Brian Stabinger, helped build western Lake Superior's contemporary surf culture. Here, Tema exits the water at Stoney Point, a two-way reef about twelve miles outside of Duluth. Photo used with permission

Tema and Stabinger to Michigan's Upper Peninsula in 1999, about a year after I made the graduate-school move to Minnesota. The second I hit the water I knew my 5-mm wetsuit, squid lid, and Northern California booties were totally inadequate for the painfully cold lefts at Union Bay.

With global heating, however, winter temperatures have been moderating. In Duluth, average December temperatures rose 6.4 degrees Fahrenheit between 1970 and 2022, and, in Northeast Minnesota more broadly, the average annual temperature increased by 3.5 degrees Fahrenheit between 1895 and 2023, with most of this warming occurring during the coldest months.[26] Moreover, the average annual temperature, according to the University of Minnesota Climate Adaptation Partnership, was projected to increase another 3.7 to 4.4 degrees Fahrenheit by the middle of the twenty-first century.[27]

Even with this long-term warming, winters can still feel bracingly cold by non-arctic standards, with actual temperatures periodically falling below zero and "feels-like" temperatures, which account for wind and other variables, hovering there much more often. Fortunately for land-locked surfers, advances in wetsuit technology have made winter surfing on the Great Lakes possible.[28] And probably no place has received more surfing acclaim in recent years than Duluth. The area's most famous break, Stoney Point, has been featured in several documentary films and a lengthy *Surfer* magazine spread.[29] Video footage of Duluth's cold-water surfers is periodically aired by national media outlets, and the local surf community has been covered by print publications such as the *New York Times*.[30] Even higher-education institutions have contributed to the region's surf-based reputation. The Recreational Sports Outdoor Program at the University of Minnesota, Duluth, offers students various surfing opportunities, and the campus created a "surf shack" at the city's most popular pointbreak. Moreover—and, full disclosure here, I am the instructor for both of these—students at the university can take an Honors surfing class or join its biennial surfing-based study-abroad program in Central America.[31]

In fact, Duluth's place in the climate crisis—a mid-sized city with abundant fresh water that has remained relatively cool on a rapidly heating planet—has received widespread attention for reasons far beyond its community of surfers. Recognizing economic potential in the worldwide environmental changes wrought by industrial emissions, some boosters began marketing Duluth as a "refuge" for "climate migrants" in the years leading up to the COVID-19 pandemic. (With its ample space, rich cultural life, and abundant outdoor opportunities, Duluth also emerged as a popular destination for those able to work from home during the pandemic.)

Duluth is "the most climate-proof city in America," Jesse Keenan of the Harvard University Graduate School of Design declared in 2019, half in jest, to a packed house at the University of Minnesota, Duluth. He even mocked up a number of prospective advertisements for the city, including two that featured a surfer silhouetted against a colorful sky and the slogan "Duluth: Not as Cold as You Think." "We found out there's a killer surfing scene here," Keenan told his local audience, including the city's then-mayor, Emily Larson. But "[w]hat we didn't tell the people in Florida is this happens in the winter only."[32]

The idea of Duluth as a "climate refuge" has achieved a surprising level of global publicity, including through reports in the *New York Times*, CNN, and *The Independent*; in podcasts for *Foreign Policy* and *The New Republic*; and, in a six-and-a-half-minute comical piece, on *The Daily Show* with Michael Kosta.[33] One of the articles in the *New York Times* (there have been two), noting how hundreds of new residents have made the move to Duluth since Keenan's 2019 lecture, opened with the story of surfer John Jenkins, a "child of Orange County" who, a decade earlier, migrated to the city to "escape overpopulation and intensifying wildfires" in California. "Surfing in Minnesota is just like surfing in Southern California, John Jenkins says, as long as you ignore the icicles on your wetsuit," the article somewhat predictably opened (Fig. 5.3).

The piece was accompanied by a striking yet obviously staged photograph of a smiling, wetsuit-clad Jenkins, board in hand, standing on ice under a blue sky along the lakeshore and appearing as if he just got out of the water—with, it should be noted, nary a wave in sight.[34]

Keenan has been careful to note that there are different categories of climate migrants. There are displaced people—many of them from the Global South—who are being forced to move by climate-related disasters, such as hurricanes, floods, droughts, or wildfires. Most of them are poor and desperate and lack the financial resources to choose to relocate. Then "[t]here is a cohort of Americans who have the means, the resources, and they have the elective mobility to be on the move," Keenan notes.[35] These, by and large, are the migrants starting to make their way to

Fig. 5.3 John Jenkins making his way across a lakeshore ice shelf after supposedly surfing the then-waveless waters of Lake Superior. Photo by Jenn Ackerman used with permission

Duluth. Surfers, who as coastal residents in places such as California generally enjoy higher incomes than their inland contemporaries, would likely fall within this latter category. Yet as even some of these "climate refugees" point out, there is no genuine refuge from the climate crisis, so at a time of rising sea levels and ferocious storm systems devastating coastlines worldwide, there are no surf zones—whether in fresh water or salt—that will escape the effects of the heating planet.[36]

What will this mean for Duluth and Lake Superior? The reality is that we don't know. But we can already see how the climate crisis is beginning to change local conditions, both terrestrial and aquatic. With the effects of the climate crisis becoming increasingly intolerable elsewhere, the expected arrival of "climate refugees" like Jenkins from more obviously climate-stricken portions of the country, such as California and Florida, may stress crowd sizes in the surf zone. This matters because a small increase in the number of surfers can have an outsize effect on the freshwater surfing experience. Unlike on an ocean, the swell window on Lake Superior (and on other lakes, from Tahoe to Michigan) is usually measured in hours, not days. With a more limited period during which it's possible to surf, and with decent waves usually at only a handful of local breaks—the short swell window makes travel to more distant spots difficult—there is greater competition for waves. The addition of just a handful of people can thus seriously impede existing surfers' ability to enjoy the freshwater experience.

The ecological changes afflicting Northeast Minnesota are not inconsequential. Some of these more directly bear on surfers' terrestrial quality of life, including the area's habitability. The once abundant moose population in the region has declined by approximately 60 percent since 2009, for example, with moose a critical subsistence resource for the Grand Portage Band of Lake Superior Chippewa, whose reservation along Lake Superior's north shore extends roughly eighteen miles from the U.S.-Canada border.[37] Warmer, drier summers will likely increase the incidence of wildfires, with the smoke from these fires leading to and exacerbating diseases such as asthma, bronchitis, heart attack, and cancer.[38] And the availability of popular local food sources, such as the region's storied wild rice, may decline "because of increasing spring precipitation and little snowfall in the winter."[39]

While such changes may negatively affect surfers' everyday experiences as residents of the Great Lakes area, surf conditions are also likely to see significant changes. One anticipated feature of the climate future—and in fact one that the region has already been witnessing—is alternating periods of drought and excessive rainfall. This will affect lake levels, as I address below, but it will also matter for water quality. Heavy precipitation will not only cause rivers and creeks to pour unusually large deposits of sediment into the Great Lakes, with probable implications for local surf zones, but also to push many hazardous pollutants into the nearshore area. The Great Lakes, which span eight U.S. states and two countries, are largely surrounded by cities, factories, and farms. (Lake Superior is an exception.) Torrential downpours induced by the heating planet thus have the potential to flood surf zones with sewage, industrial waste, and agricultural runoff. In the same way that ocean-based surfers have historically been advised to avoid the filthy lineups of certain surf breaks following large storms, Great Lakes surfers may have to increasingly calculate the risks associated with pursuing their pastime in polluted local waters.

Yet it's not just about water quality. The surface water temperatures of Lake Superior have been warming even faster than the rapidly warming air temperatures, with ice coverage—both the amount and its duration—decreasing over time.[40] This may not sound like a negative outcome for surfers, who require open water and usually appreciate warmer rather than colder temperatures, but the surf-related effects of this development may be considerable. Waves on Lake Superior are significantly larger in the colder months, so while the decrease in ice coverage may mean more opportunities to take advantage of wave activity, warmer air and water temperatures and the lack of ice may prove detrimental to existing surf breaks and conditions. For one thing, colder water is often associated with stronger wind patterns, and colder, denser air creates larger waves than warmer air; warmer water and air temperatures may thus affect the generation of Lake Superior wind swell. But that's not all. Ice acts as a protective barrier to shorelines (as well as providing sanctuary to aquatic life), so an absence of ice will likely result in greater erosion and coastal destruction.

The effects of this reduced ice cover may be exacerbated by greater variability in lake levels. Lake Superior does not have tides, and thus freshwater surfers, unlike those on the oceans, do not have to factor tidal changes into their decisions about where and when to surf. (Lake Superior does see surges and seiches, however.) Yet in the same way that oceans are experiencing rising sea levels from melting ice in the polar regions, both droughts and more intense storms, such as the five to ten inches of rain that fell on Duluth from June 19–20, 2012—famously leading to a globally-circulated photograph of Feisty the seal on a nearby street after floods swept her from her zoo enclosure[41]—will likely result in greater variability in lake levels and thus, during periods of high water, potentially greater susceptibility to erosion and coastal damage.[42]

In other words, there may be periods that seem like an endless high tide and there may be periods that seem like an endless low tide. This will affect surf conditions, as water levels affect the quality of lake breaks in the same way that tidal changes affect the quality of ocean breaks, and these water levels may significantly affect the health of the surf zone as well. For example, water levels in Lake Superior rose considerably to monthly record highs in 2019 and 2020.[43] Not only did this result in flooding and wave damage to lakeside homes and public infrastructure in communities such as Duluth, but the elevated levels appeared to affect the quality of the surf at several local breaks. At both The Rock, which is just north of the mouth of Duluth's Lester River, and Stoney Point, which is about twelve miles outside the city and is by universal acclaim the premier wave on the Great Lakes, this persistent "high tide" appeared, at least anecdotally, to result in fewer surfable days and mushier conditions.

Higher or lower water levels also have the potential to permanently alter the Lake Superior nearshore topography—and thus its surf zones. It's true that new surf breaks may emerge as greater susceptibility to erosive wave activity or low water levels alter the Superior shoreline, but a number of existing surf breaks are likely to be degraded or lost. Just in the last decade surfers in Duluth have twice seen significant shoreline alterations at a spot called Southside, one of the most popular surf breaks in the city. Southside is a cobblestone left point that sits adjacent to the mouth of the Lester River and within easy walking or paddling distance of The Rock. Large storms with heavy rainfall pushed so many stones either down the river or along the shoreline that a new rocky outcrop—one that for a brief spell even contained a lagoon—formed at the northeast corner of the point. The two storm events happened years apart, and in both cases months of wave activity ultimately broke the rock extensions up, with the stones presumably pushed farther down the shore. But while the outcrops existed, they seemed to interfere with the quality of the surf. They did not discernibly extend the length of the wave or improve surfing conditions. On the contrary, the rock outcrops appear to have degraded what had previously been a reasonably consistent and reliably long left. While I cannot say with certainty that the rock deposits were the cause of this degradation, I can say as someone who has been surfing Southside for approximately twenty years that I noticed a diminution in wave quality and length of rides over the last decade.

Freshwater surf zones may also be harmed by the heating planet's potential to cause a greater incidence of "meteotsunamis," or meteorological tsunamis. These phenomena occur when a rapid rise in water levels from an overland storm that has moved over a body of water hits a shallow nearshore shelf and produces a tsunami-like "wave" that, on the Great Lakes at least, can potentially be quite large. Most are small—two feet or less—but researchers from the University of Wisconsin-Madison believe that several historical tragedies previously attributed to seiches were in fact caused by meteotsunamis, including a 20-foot wave that surged over a pier in Grand Haven, Michigan, in 1929, killing ten people, and a 10-foot wave that hit Holland, Michigan, in 1938, killing five others.[44] Lake Superior, together with Lake Ontario, has seen the fewest number of meteotsunamis, while Lake Michigan has seen the greatest. But across the region the outlook appears troubling. "Future climate scenarios over the United States show a likely increase in the number of days favorable to severe convective storm formation over the Great Lakes, particularly in the spring season," a Wisconsin-led team of scientists found. "This would suggest that the convectively associated meteotsunamis in these regions may experience an increase in occurrence frequency or a temporal shift in occurrence to earlier in the warm season."[45]

And then there is a biological threat, with rising water temperatures resulting in a likely increase in the incidence of algal blooms. A major bloom first appeared on Lake Superior in 2012, followed by another large bloom in 2018. These may have been unsightly and unpleasant, but they were minimally harmful. Then toxins associated with blue-green blooms were discovered in Lake Superior—though at a protected beach in the Duluth-Superior harbor—for the first time in 2021.[46] While scientists have publicly expressed skepticism that such toxins could thrive in the deeper, colder, and more turbulent waters of the open lake, it is not inconceivable that, with warming surface temperatures, Superior surfers may one day have to contend with harmful algal blooms in the same way that Lake Erie sees them or that ocean surfers are occasionally counseled to avoid "red tides."[47]

*****

Precisely what the future holds for Lake Superior, the Great Lakes more broadly, or other non-oceanic surf zones is uncertain. With portions of the industrialized world, led by a climate crisis-denying right-wing in the United States, insistent on continuing to generate harmful greenhouse gases ("drill, baby, drill," as two-time president Donald Trump, following the example of earlier Republican officials, has repeatedly put it), a very different existence is coming.[48] Perhaps the only thing we can be certain of is that the surfing experience of the future will look quite different from that of the past. Melting polar ice will affect sea levels on the world's oceans, while seasons punctuated by drought or excessive precipitation will affect lake levels and river flows. But it's not just about where waterlines might find themselves. The heating planet will affect just about every facet of human existence, including the most fundamental one of all: Will humans continue to reproduce in numbers sufficient to perpetuate the species? Already many young people are announcing their decisions not to do so. Those people who remain are likely to see greater human migration to so-called "climate refuges" as continued residency in areas afflicted by scorching temperatures, massive wildfires, and more intense hurricanes seems less and less viable.

The question of what might become of surfing may seem frivolous in the face of these much greater existential challenges. After all, how important, really, are the pleasures associated with wave-riding when millions or billions of people are struggling to merely survive? Not very. Yet, as countless surfers can attest, the pleasure that people find in the waves is what often makes life worth living for them. There is a reason that so many people have built their lives around their ability to surf. That is unlikely to change, though what form it takes might. As the late twentieth and twenty-first centuries have witnessed surfing spread beyond the world's oceans to its lakes, rivers, and artificial pools, the future of such unconventional waves—like the future of saltwater surfing as well—remains unknown. But it most certainly will look different.

Notes

1. Juan Hernandez, "WSL Adds KSWC's Surf Abu Dhabi to the'25 CT Schedule," *The Inertia*, August 15, 2024, at <www.theinertia.com/surf/wsl-announces-kswcs-abu-dhabi-wave-pool-will-be-on-the-25-ct-schedule/> (accessed October 30, 2024).
2. Human Dignity Trust, "United Arab Emirates," October 26, 2023, at <www.humandignitytrust.org/country-profile/united-arab-emirates/> (accessed October 30, 2024).
3. Amnesty International, "United Arab Emirates (UAE)," n.d., at <www.amnestyusa.org/countries/uae/> (accessed October 30, 2024).
4. Alexander Haro, "Tyler Wright Takes a Knee in Solidarity with Black Lives Matter at Tweed Coast Pro Event," *The Inertia*, September 13, 2020, at <www.theinertia.com/surf/tyler-wright-black-lives-matter-takes-knee-tweed-coast-pro-surfing-event/> (accessed October 30, 2024); Cooper Gegan, "WSL Criticized Over Championship Tour Stop's Anti-LGBTQIA + Policies," *The Inertia*, October 16, 2024, at <www.theinertia.com/uncategorized/wsl-criticized-over-championship-tour-stops-anti-lgbtqia-policies/> (accessed October 30, 2024).
5. Gegan, "WSL Criticized Over Championship Tour Stop's Anti-LGBTQIA + Policies"; Evan Quarnstrom, "Tyler Wright's Family Skewers WSL for Competing Under UAE's Anti-LGBT Laws; We Spoke to a Local," *The Inertia*, October 16, 2024, at <www.theinertia.com/surf/tyler-wrights-family-skewers-wsl-for-competing-under-uaes-anti-lgbt-laws-we-spoke-to-a-local/> (accessed October 30, 2024).
6. water_lilli, Instagram, October 13, 2024, at <www.instagram.com/p/DBD4hL0Tv7O/?img_index=1> (accessed October 30, 2024). See also Quarnstrom, "Tyler Wright's Family Skewers WSL for Competing Under UAE's Anti-LGBT Laws."
7. Holden Trnka, "Confirmed: Abu Dhabi Wavepool to Appear on 2025 CT," *Stab*, August 15, 2024, at <stabmag.com/elsewhere/confirmed-abu-dhabi-wavepool-to-appear-on-2024-ct/> (accessed October 30, 2024); emphasis in the original. For a brief history of wave pools, including twentieth-century wave-pool contests, see Matt Warshaw, *The Encyclopedia of Surfing* (Orlando, Florida: Harcourt, Inc., 2003), 683–684.
8. Ismaeel Naar and Vivian Yee, "A March Forward, or a Mirage?" *New York Times*, November 12, 2024. Indeed, Saudi Arabia won the bid to host the 2034 men's FIFA World Cup following an unprecedented and controversial voting arrangement. Tariq Panja, "FIFA Bends Own Rules to Give Saudi Arabia Coveted 2034 World Cup," *New York Times*, December 10, 2024, at <www.nytimes.com/2024/12/10/world/middleeast/saudi-world-cup-human-rights.html> (accessed December 10, 2024). On the response of human rights organizations to this decision, see "Joint Statement: Award of 2034 World Cup to Saudi Arabia Risks Lives and Exposes FIFA's Empty Human Rights Commitments," December 11, 2024, at <www.amnesty.org/en/documents/IOR10/8832/2024/en/> (accessed December 18, 2024).
9. See, for example, Patrick Jack, "Scholars Blast NYU Abu Dhabi 'Crackdown' on Palestinian Support," *Inside Higher Ed*, June 28, 2024, at <www.insidehighered.com/news/global/us-colleges-world/2024/06/28/scholars-blast-nyu-abu-dhabi-crackdown-palestinian-support> (accessed November 14, 2024).
10. "The Louvre Abu Dhabi," n.d., at <www.louvre.fr/en/the-louvre-in-france-and-around-the-world/the-louvre-abu-dhabi> (accessed November 14, 2024).

11. Trnka, "Confirmed: Abu Dhabi Wavepool to Appear on 2025 CT." Surfrider Europe, the European arm of the Surfrider Foundation, registered its opposition to wave pools in 2020 because they require "land artificialization," "huge water consumption," and "a large quantity of energy." Surfrider Foundation Europe, "Wave Pools: Environmental Concerns Outweigh Their Value," June 15, 2020, at <www.surfrider.eu/learn/news/wave-pools-environmental-concerns-outweigh-their-value/> (accessed December 19, 2024).
12. Matt Hapgood, "Why Aren't There More Saltwater Wave Pools?" *WavePoolMag*, August 4, 2024, at <wavepoolmag.com/why-arent-there-more-saltwater-wave-pools/> (accessed October 30, 2024).
13. Kurt Badenhausen, "Facebook and World Surf League Announce Exclusive Partnership," *Forbes*, January 24, 2018, at <www.forbes.com/sites/kurtbadenhausen/2018/01/24/facebook-and-world-surf-league-announce-exclusive-partnership/> (accessed November 14, 2024); Eric Fisher, "NBA, Real Madrid Tops in Social Engagement," *Sports Business Journal*, July 17, 2017, at <www.sportsbusinessjournal.com/Journal/Issues/2017/07/17/Research-and-Ratings/Social.aspx> (accessed November 14, 2024). That statistic may have been skewed somewhat by the fact that a video the WSL posted of dolphins surfing near a contest in Western Australia was "the most-watched video globally over the past year, drawing more than 121.4 million views." David Broughton, "Data Shows World Surf League Had Most-Watched Video Globally Over Last Year," *Sports Business Journal*, July 21, 2017, at <www.sportsbusinessjournal.com/Daily/Issues/2017/07/21/Research-and-Ratings/Hookit.aspx> (accessed November 14, 2024).
14. Cooper Gegan, "American Wave Machines to Build China's First Wave Pool," *The Inertia*, December 17, 2024, at <www.theinertia.com/surf/american-wave-machines-china-first-wave-pool/> (accessed December 19, 2024).
15. Weird Waves, at <www.youtube.com/playlist?list=PLEfn8qMSGQa8kVZtBtIWfH_3wU_iE7vJQ> (accessed December 19, 2024).
16. See, for example, the videos of YouTube personality Ben Gravy, who has, among other things, surfed a wave pool in South Korea, dropped into river waves in West Virginia, Kentucky, and Tennessee, and paddled out in Lake Superior. Ben Gravy, at <www.youtube.com/channel/UC4qa5EroJxnQM6Sng9WhgsQ> (accessed December 19, 2024).
17. Cooper Gegan, "The Definitive List of Man-Made River Surfing Waves," *The Inertia*, August 6, 2024, at <www.theinertia.com/surf/the-definitive-list-of-man-made-river-surfing-waves/> (accessed December 19, 2024).
18. For Great Lakes surface water temperatures, see Fig. 2 in "Climate Change Indicators: Great Lakes Water Levels and Temperatures," United States Environmental Protection Agency, June 2024, at <www.epa.gov/climate-indicators/great-lakes> (accessed December 18, 2024).
19. "Lake Superior," Great Lakes Commission, no date, at <www.glc.org/lakes/lake-superior> (accessed December 18, 2024).
20. William Povletich, "Malibu of the Midwest," *Wisconsin Magazine of History* 96:1 (Autumn 2012): 46–47.
21. Povletich, "Malibu of the Midwest," 41; Rick Boss, "Great Lakes, Great Surfing," *active magazine* [Marquette, Michigan] (Summer 1999): 9.
22. Lake Superior is surpassed only by Lakes Baikal and Tanganyika in freshwater volume. For some of the attention Lake Superior surfing has recently received, see, for example, *Unsalted: A Great Lakes Experience* (dir. Vince Deur, 2005); Red Bull, "Surfing in the Great Lakes – Red Bull New Wave," April 7, 2011, at <www.youtube.com/watch?v=6sMWoG0llYo> (accessed November 26, 2024); Justin Housman, "The Other North Shore," *Surfer*, August 23, 2016, at <www.surfer.com/news/the-other-north-shore-great-lakes> (accessed November 26, 2024); Vans, "Weird Waves Season 1: Great Lakes | Surf | VANS," January 8, 2019, at <www.youtube.com/watch?v=8DmoQ-ziT38> (accessed November 26, 2024); Aloha Visuals, "Surfing Lake Superior with Alex Gray," October 4, 2021, at <www.youtube.com/watch?v=_ZNmr8U0Dbw> (accessed November 26, 2024); and Ben Gravy, "Swell of the Century in Minnesota," January 9, 2022, at <www.youtube.com/watch?v=0cKIDYfOubM> (accessed November 26, 2024).

23. Povletich, "Malibu of the Midwest," 45; Bradley Campbell, "Great Lakes Surfers Pay No Mind to Frigid Temps," *City Pages* [Minneapolis] (December 10, 2008).
24. *The Surf Report* 14:1 (January 1993): 1–4. Minnesota's breaks "are remote and surfing on Lake Superior should be done with caution," the publication noted. "The water temperature rarely reaches 60°F and only for 3–5 weeks, usually in August. Lake Superior is the largest freshwater lake in the world and has been known to generate hurricane-force winds. These conditions are noted by the U.S. Coast Guard and most locals remember November 10, 1975[,] when the iron-ore carrying Edmund Fitzgerald sank in 20–30 foot seas and hurricane force winds in excess of 80 MPH off Whitefish Bay near Sault Ste Marie, Michigan." Ibid., 1.
25. Lisa Rounds, "Subzero Surfing on Lake Superior," *Live / Play / AAA* (January/February 2014): 29.
26. John Myers, "Record Warm Duluth December Part of Long-Term Trend," *Duluth News Tribune*, January 1, 2024, at <www.duluthnewstribune.com/news/local/record-warm-duluth-december-part-of-long-term-trend> (accessed December 2, 2024); University of Minnesota Climate Adaptation Partnership, "Climate Change in Northeast Minnesota," September 2024, at <https://climate.umn.edu/climate-change-northeast-minnesota> (accessed December 2, 2024).
27. University of Minnesota Climate Adaptation Partnership, "Climate Change in Northeast Minnesota," September 2024.
28. Ironically, however, the manufacture of most wetsuits – nearly all are made from neoprene, which is a petroleum-derived synthetic rubber – has contributed to the climate crisis. An exception are the wetsuits first manufactured by Patagonia; they are made mostly from Yulex, a natural rubber tapped from hevea trees.
29. Justin Housman, "The Other North Shore," *Surfer*, August 23, 2016, at <www.surfer.com/news/the-other-north-shore-great-lakes> (accessed December 30, 2024). On Stoney Point's appearance in documentary films, see, for example, *Unsalted: A Great Lakes Experience* (dir. Vince Deur, 2005) and *Freshwater* (dir. Ian Planchon, 2022).
30. "Minnesota Surfers Shred Big Waves on Lake Superior," ABC News, May 1, 2017, at <www.facebook.com/ABCNews/videos/minnesota-surfers-shred-big-waves-on-lake-superior/10155706525503812/> (accessed December 30, 2024); "Surfers Brave Frigid Temperatures Catching Waves on Lake Superior," NBC News, February 23, 2023, at <www.nbcnews.com/video/surfers-brave-frigid-temperatures-catching-waves-on-lake-superior-163950661634> (accessed December 30, 2024); "Minnesota's Lake Superior Becomes Hot Spot for Winter Surfing," CBS Evening News, February 25, 2023, at <www.youtube.com/watch?v=BgBamHIK1SA> (accessed December 30, 2024); Stephen Regenold, "Hanging 10 (Degrees) on Icy Lake Superior," *New York Times*, January 15, 2009, at <www.nytimes.com/2009/01/16/travel/escapes/16superior.html> (accessed December 30, 2024), with accompanying photographs by T.C. Worley at <www.nytimes.com/slideshow/2009/01/15/travel/escapes/0116-superior_index.html> (accessed December 30, 2024).
31. On the "surf shack," see "Surf Shack," no date, at <https://umdrsop.d.umn.edu/facilities/surf-shack> (accessed December 30, 2024).
32. Jesse Keenan, "Destination Duluth: The Fact and Fiction of a Shared Climate Future," University of Minnesota, Duluth, April 15, 2019, at <www.youtube.com/watch?v=vUhSckQsrwY> (accessed November 25, 2024). Keenan has since moved to Tulane University.
33. Kendra Pierre-Louis, "As the Earth Warms, Duluth Is Looking Pretty Cool," *New York Times*, April 16, 2019; Debra Kamin, "Out-of-Towners Head to 'Climate-Proof Duluth,'"

New York Times, March 13, 2023, at <www.nytimes.com/2023/03/10/realestate/duluth-minnesota-climate-change.html> (accessed November 25, 2024); John D. Sutter, "As People Flee Climate Change on the Coasts, This Midwest City Is Trying to Become a Safe Haven," CNN, April 12, 2021, at <www.cnn.com/2021/04/12/opinions/climate-migration-in-america-california-duluth-sutter/index.html> (accessed November 25, 2024); Sebastien Malo, "US Cities Preparing to Become Havens for Mass Migration from Climate Change," *The Independent*, April 6, 2019, at <www.independent.co.uk/news/world/americas/climate-change-us-migration-ohio-new-york-global-warming-a8857636.html> (accessed November 25, 2024); "Climate Migrants: Destination Duluth," *Foreign Policy*, November 24, 2021, at <https://foreignpolicy.com/podcasts/heat-of-the-moment-climate-change/climate-migrants/> (accessed November 25, 2024); Alex Pareene and Laura Marsh, "In Pursuit of the Climate-Proof City," *The New Republic*, April 12, 2023, at <https://newrepublic.com/article/171759/pursuit-climate-proof-city> (accessed November 25, 2024); and "Is Duluth, MN a Climate Change Safe Haven? Michael Kosta Investigates," *The Daily Show*, February 22, 2024, at <www.youtube.com/watch?v=83DX5XDwz40> (accessed November 25, 2024).

34. Kamin, "Out-of-Towners Head to 'Climate-Proof Duluth.'".
35. Dan Kraker, "Climate-Proof Duluth? Why the City Is Attracting 'Climate Migrants,'" Minnesota Public Radio, October 4, 2021, at <www.mprnews.org/story/2021/10/04/climateproof-duluth-why-the-city-is-attracting-climate-migrants> (accessed November 25, 2024).
36. See, for example, the comments of transplanted Californian Jamie Alexander in "Is Duluth, MN a Climate Change Safe Haven? Michael Kosta Investigates."
37. MPR News Staff, "Minnesota Moose Count Falls in Latest Survey; DNR Says Trend Stable," MPR News, April 6, 2023, at <www.mprnews.org/story/2023/04/06/minnesota-moose-count-falls-in-latest-survey-dnr-sees-stable-trend> (accessed December 2, 2024); Minnesota Indian Affairs Council, "Gichi-Onigaming / Grand Portage Band of Lake Superior Chippewa," n.d., at <https://mn.gov/indian-affairs/tribal-nations-in-minnesota/gichi-onigaming-grand-portage-band-of-lake-superior-chippewa.jsp> (accessed December 3, 2024).
38. University of Minnesota Climate Adaptation Partnership, "Climate Change in Northeast Minnesota."
39. University of Minnesota Climate Adaptation Partnership, "Climate Change in Northeast Minnesota."
40. Jay A. Austin and Steven M. Colman, "Lake Superior Summer Water Temperatures Are Increasing More Rapidly Than Regional Air Temperatures: A Positive Ice-Albedo Feedback," *Geophysical Research Letters* 34:6 (2007): 1–5; Stephanie E. Hampton, et al., "Environmental and Societal Consequences of Winter Ice Loss from Lakes," *Science* 386:162 (October 11, 2024): 1–10. See also Jimmy Lovrien, "Minnesota Lakes Have Lost 2 Weeks of Ice Cover in 50 Years," *Duluth News Tribune*, November 29, 2024, at <www.duluthnewstribune.com/news/local/minnesota-lakes-have-lost-2-weeks-of-ice-cover-in-50-years> (accessed December 2, 2024).
41. Madeleine Baran, "Eyewitnesses Describe Frantic Search for Loose Duluth Zoo Animals," Minnesota Public Radio, June 22, 2012, at <www.mprnews.org/story/2012/06/22/eyewitnesses-describe-frantic-search-for-loose-duluth-zoo-animals> (accessed December 2, 2024).
42. Jimmy Lovrien, "Experts: Climate Change Driving Additional, More Intense Rain in Minnesota," *Duluth News Tribune*, July 12, 2024, at <www.duluthnewstribune.com/news/local/experts-climate-change-driving-additional-more-intense-rain-in-minnesota> (accessed July 12, 2024).
43. John Myers, "Lake Superior Missed All-Time Water Level Record by 1.1 Inches," *Duluth News Tribune*, November 4, 2019, at <www.duluthnewstribune.com/news/lake-superior-missed-all-time-water-level-record-by-1-1-inches> (accessed December 18, 2024); John Myers, "Lake Superior Water Levels Nearing Normal," *Duluth News Tribune*, June 3, 2021, at <www.duluthnewstribune.com/news/lake-superior-water-levels-nearing-normal> (accessed December 18, 2024). They have since dropped to closer to their long-term average.

44. John Myers, "Researchers Discover 'Tsunamis' on Great Lakes," *Duluth News Tribune*, April 25, 2016, at <www.duluthnewstribune.com/news/researchers-discover-tsunamis-on-great-lakes> (accessed December 18, 2024).
45. Adam J. Bechle, et al., "Meteotsunamis in the Laurentian Great Lakes," *Scientific Reports* 6:37,832 (November 24, 2016): 1. See also Marie Zhuikov, "Researchers Find Meteotsunamis Are an Underrated Hazard in the Great Lakes," University of Wisconsin Sea Grant, November 28, 2016, at <www.seagrant.wisc.edu/news/researchers-find-meteotsunamis-are-an-underrated-hazard-in-the-great-lakes/> (accessed December 18, 2024).
46. Danielle Kaeding, "For the First Time, Toxins Found in Blue-Green Algae Bloom in Inlet of Lake Superior," Wisconsin Public Radio, September 30, 2021, at <www.wpr.org/environment/first-time-toxins-found-blue-green-algae-bloom-inlet-lake-superior> (accessed December 2, 2024).
47. On Lake Erie, see "Lake Erie Harmful Algal Bloom Forecast," National Centers for Coastal Ocean Science, no date, at <https://coastalscience.noaa.gov/science-areas/habs/hab-forecasts/lake-erie/> (accessed December 19, 2024). For surfing and "red tides," see Dashel Pierson, "Red Tide: Everything You Need to Know," *Surfline*, May 10, 2018, at <www.surfline.com/surf-news/red-tide-everything-need-know/23478> (accessed December 19, 2024).
48. Watcher.Guru, X, May 10, 2023, at <https://x.com/WatcherGuru/status/1656462087055368195> (accessed December 26, 2024).

Chapter 6
Wave Saving Technology: Harnessing Innovation for Surf Ecosystem Conservation

Diego A. Sancho-Gallegos

Introduction

This book explores the future of technology and surfing. At the onset, I have to say, it does not matter if I am a pessimist or an optimist about technology. Technology is going to continue moving forward, regardless of whether I agree with the social, cultural, and environmental costs associated with its rapid development or not. Therefore, I like to be on the side of trying to do something productive with it, harnessing innovation to better steward our environment—and yes, our waves too.

This chapter will focus on the different applications of technology in surf ecosystem conservation; from success stories to the challenges we constantly face. I hope it'll inspire skeptics and believers alike, and shine a light into the possibilities the future holds. At the end of the day, technology is a tool and, like any other tool, it only matters how we develop it and use it–it is not inherently good or bad.

In my role at Save The Waves Coalition, I have spent several years managing the Save The Waves App and leading our data-driven stewardship and resilience efforts, which places me at the forefront of utilizing technology for effective surf ecosystem conservation around the world. For context, Save The Waves Coalition (STW) is an international nonprofit dedicated to protecting surf ecosystems across the globe in coalition with diverse organizations. Our vision is to facilitate a world where surf ecosystems are valued and where surfing provides a proactive vehicle for long-term coastal conservation.

Surf ecosystem conservation can only be done by people coming together to operate within and around the spaces that need protection, what we call the surf ecosystem—the land-to-sea interface where waves for surfing occur, plants and animals live and thrive, and human communities recreate and make a living. Our goal is to empower people with the resources they need to implement and execute

D. A. Sancho-Gallegos (✉)
Save The Waves Coalition, 849 Almar Ave, Suite C#163, Santa Cruz 95060, CA, USA
e-mail: diego@savethewaves.org

L. Mach (ed.), *The Future of Surfing in the Anthropocene*,
https://doi.org/10.1007/978-3-032-09377-6_6

projects on the ground in a variety of diverse contexts around the world. That is why we are a coalition.

Our global coalition of advocates, partners, and supporters drives our work. However, as most environmental organizations, we find ourselves with limited resources to constantly monitor all the data and information about threats coming in from, literally, all over the world. We are honored to be a resource to help coastal enthusiasts protect the places they love, but as threats continue to intensify and spread around the world, we really need innovative solutions that enhance how we do coastal stewardship. That's where our goal of using technology to facilitate timely information transfer and coalition building stems from. We want to help communities organize, respond to, and prevent threats effectively. By leveraging technology, we are able to empower everyone, regardless of where they are in the world, to engage with coastal issues in their communities and work together to find solutions.

Before I go further, let me give you a brief overview about how I got here. I was born and raised in Costa Rica, so I have always experienced surfing as a growing and important part of coastal economies. I witnessed a lot of coastal towns changing very quickly after surfers started to "discover" these places and tourism businesses were established to cater to surfers. Like most human activities, this was a complex and nuanced change—some parts were positive and some were negative, and depending on who you would ask (local surfers, foreign surfers, government officials, developers, etc.) they would give you a different answer. In my studies, I focused on learning about the coastal ocean and how we've attempted to manage these complex interactions. As I was completing my Master's degree, I crossed paths with Save The Waves, and that's where my journey began.

The Rise of Satellite Data: The Surf Conservation Index

My first foray into surf ecosystem conservation with Save The Waves focused on furthering the existing framework for the Surf Conservation Index (SCI). This study maps out surf spots and their characteristics to areas with high biodiversity, high economic reliance on surfing, and significant anthropogenic pressure based mostly on satellite-derived data. The SCI helps identify potential sites for surf ecosystem conservation and provides a prioritization of surf spots based on the highest conservation need and potential (Arroyo et al., 2019). The index considers five main inputs to map out conservation priorities based on: (1) pressure (i.e., population, infrastructure, land use); (2) biodiversity; (3) surf resources (i.e., how many surf breaks are there, quality, consistency, experience levels accommodated); (4) social (i.e., recreational services and industries related to surf visitation); and (5) response (i.e., existing protected area designations or conservation efforts). The full report can be found on the Save The Waves website as part of our Surf Protected Area efforts in Costa Rica (Sancho-Gallegos & Arroyo Rodriguez, 2021) (Fig. 6.1).

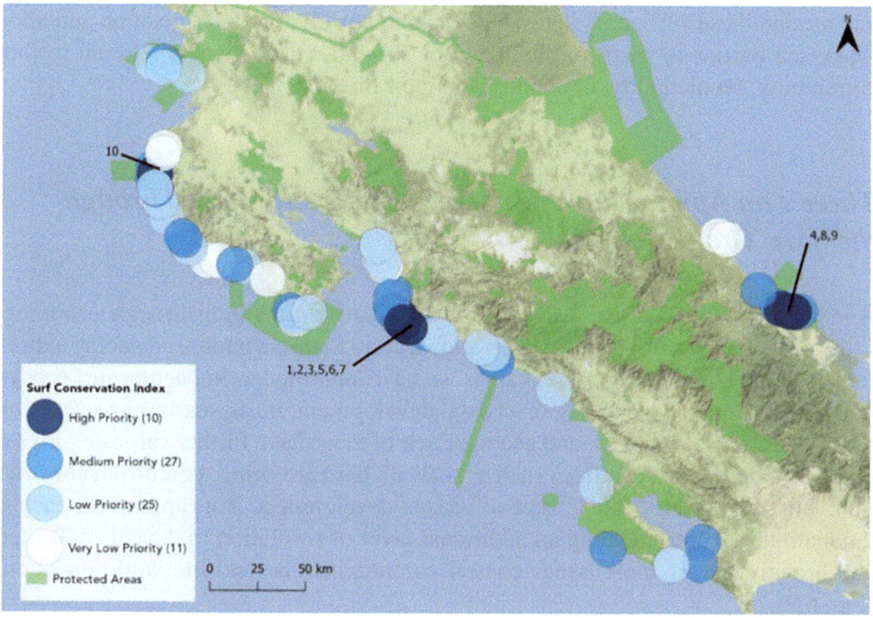

Fig. 6.1 Final Costa Rica Surf Conservation Index map (Sancho-Gallegos & Arroyo Rodriguez, 2021)

This example illustrates a simple use case of technology for surf ecosystem conservation. We harnessed large global data sets, largely derived from technological advancements in remote sensing, to map out surf ecosystem conservation priorities. This type of analysis helps highlight places where community economic dependence on surf resources is high, waves are of good quality, there is a great deal of biodiversity, and the population and infrastructure development hasn't yet outpaced local carrying capacities. We also continue to add new components to SCIs as we implement them in new locations and new datasets become available, most recently adding a climate index that includes the overlap with valuable carbon sinks and coastal vulnerability. The outcome of these SCIs is to provide local decision makers and authorities with a science-based resource that can be used to justify surf break protections and other conservation measures around these locations (Arroyo et al., 2023).

For most surfers, the highest priority waves for conservation are the ones that they surf most often. This means that, while the SCI prioritizes surf breaks relative to each other in a country, we do not want to discount the cultural importance of any wave. To build local coalitions, it can be effective for us to help demonstrate, through something like the SCI, where priorities align and there's interest and value in preserving not only waves, but the surrounding ecosystems, cultures, and economies. The SCI results are not a prescriptive "you must protect these breaks" approach, rather they are a tool that illustrates the ecosystem services and threats that exist in and around these surf ecosystems. They lend themselves to interpretation and allow interested parties to focus on the index or sub-index that is most relevant to their community, municipality, or country and their mission.

There's an App for that: Protecting Surf Ecosystems Through the Save The Waves App

Another approach Save The Waves uses to facilitate conservation through technology is the Save The Waves App. The App allows users to take a photo and geotag a threat falling in one the six major threat areas we have identified as impacting surf ecosystems in the Anthropocene Ocean: coastal development, trash, sea-level rise/erosion, water quality, reef impacts, and access. Each of these have further sub-categories of impacts people can select from such as seawall, hard armoring, beachfront construction, jetty, harbor, or "Other" type of coastal development. This applies for each of the main threat areas, adding an additional layer of resolution to the data collected, and broadly covers most issues coastal communities around the world encounter (Fig. 6.2).

The idea behind the App is to facilitate effective information sharing about threats to surf ecosystems with the goal of enabling action on the ground around the issues being reported. Save The Waves is only one of a few dedicated surf conservation organizations around the world, and arguably, the one with the most global focus. Other organizations we admire and collaborate with, like Surfers Against Sewage or Surfrider Foundation, have apps dedicated for their specialized purposes or have leveraged existing apps for specific campaigns, but prior to the Save The Waves App

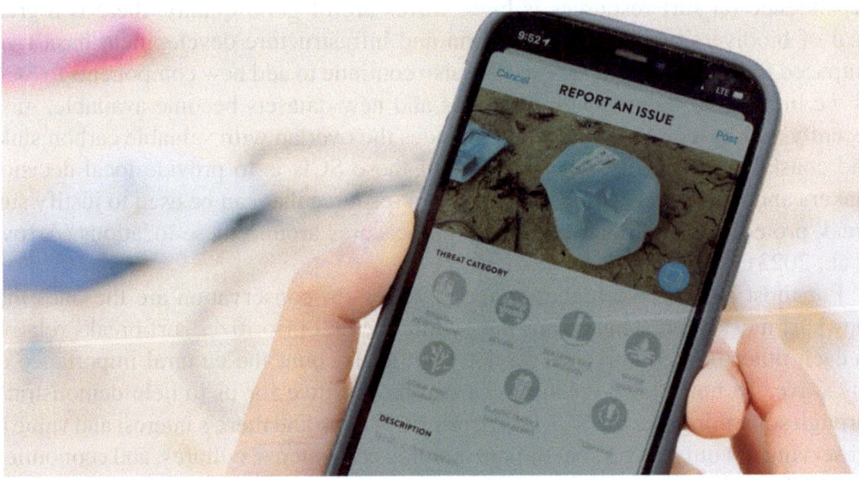

Fig. 6.2 Save The Waves App user interface

there was no generalized app any surfer could use anywhere to care for their surf ecosystems. However, as I mentioned earlier, we have limited capacity to respond to these threats around the world. Every time there is a clear threat impacting a surf ecosystem (which is occurring far too often), we receive reports from concerned surfers. We understand the urgency surfers feel, and we often share it, but we simply can't verify and respond to every threat directly.

With the App, however, we have automated a system where, based on where a threat is reported and the type of threat it is, we alert partner environmental organizations, government agencies and/or community leaders working in those areas and issues. This facilitates organizing around issues among surfers and other local stakeholders to find solutions, empowering individuals with resources and action and sparking collaboration in ways that would not have happened before.

The App also allows us to map where certain threats are occurring with the most frequency, and the type of threat, building a solid open access global database. Surfers and organizers can also utilize our interactive map, or build their own maps from our data, to showcase threats that have been reported in their region to policymakers. This can help surfers and conservationists validate challenges, gain support for their efforts, and show the extent of an issue through citizen science in new and exciting ways. The data is already being utilized across a variety of platforms and repositories to show certain types of threats and where they are emerging around the world. All App data is available through the Save The Waves App webpage (savethewaves.org/app/data), which can be freely downloaded by individuals, organizations or advocacy groups for campaigning, research projects, or local-level policy making (Fig. 6.3).

Regardless, the reality of leveraging seemingly simple technology (in our modern context) such as a mobile app is easier said than done. The initial iteration of the Save The Waves App was built through pro-bono support in 2017 (the heyday of

Fig. 6.3 The App's main threat categories

apps) when everyone was trying to adopt them for everything. This "beta" version (in tech speak) proved the model and encouraged Save The Waves to pursue this idea seriously. I personally joined Save The Waves to help roll out Version 2, the scaled-up version, which includes four different languages and an Android version, making it available for a truly global audience as more than 70% of the planet uses Android-operated devices. But, not only do you need to educate people about the existence of an app and convince them to download it, users also need to use it regularly to build muscle memory and have an incentive to continue to do so. As opposed to TikTok or Instagram or other social media apps that were designed to be addictive and keep people on them, we don't have a team of thousands of highly-paid software engineers, product managers and a huge marketing budget behind our App, which means it takes a lot more elbow grease from a few of us to make things happen. As a result, new App versions and features roll out much slower than what users of these big tech apps are used to, making engagement limited and forcing us into more creative strategies. While we haven't gamified our App the way others have, we have been able to leverage corporate partnerships and communication campaigns to reach a consistent and engaged user base.

We've concluded that if people are aware of the Save The Waves App, they will use it to tag a threat and then forget about it until they see something else threatening a surf break somewhere. That process requires identifying the threat and being willing to take action on the App, which depends heavily on users' motivation to steward their coastal ecosystems. While initially we may have had ambitions of having a huge user base of surfers, we quickly realized that it would require a significant amount of marketing budget leveraging complex behavioral science. As a result, we shifted focus towards making the App an effective tool for conservation and information sharing. We decided to ensure that information flows adequately and that it gets to the right people or organizations at the right time, finding our niche for impact with the tools we have available.

As part of this process, we learned from our experiences and have modified the App to focus more on the quality of engagement, rather than the quantity of users. This has been helpful because we identified people in parts of the world that are really trying to protect and steward places, and has allowed us to empower them to effectively do so. The App can connect motivated coastal stewards to the right resources and organizations to support strategic planning for larger projects or government engagement, which the App can also support with its data.

In essence, the App feeds into all the other Save The Waves programs, such as the World Surfing Reserves, Surf Protected Areas, and Endangered Waves programs. The idea is that labeling threats can help bring together coalitions and build partnerships that can coalesce around broader objectives for surf ecosystem conservation. We see the App as empowering local level stewardship not just for surfers, but across anyone who uses or cares about the oceans and coastlines. Because our App is free and open to all, other small organizations can utilize it to raise awareness of certain threats and build support for their efforts, creating additional coalitions for conservation beyond surfers.

Fig. 6.4 Fiji stewardship efforts led by Hannah Bennett

One example of how the App has worked was our Dirty Wave Challenge in 2018. The idea was to spread awareness about the App and provide an incentive for downloading it and tagging local threats. The issue or area that demonstrated the most imminent threat was offered a $5000 grant to put towards working to tackle the issue. This first winner was Hannah Bennett from Fiji, who is part of the Fiji Surfing Association and a passionate environmentalist and community leader. Hannah's goal was to reclaim one of the main beaches in Suva, where she grew up surfing. She had been living there her whole life, and she had seen it impacted by trash, like a lot of other places, which in turn was affecting local perception and use of this resource. She was seeing less people spending time at this beach and less youth taking up surfing as a result of the apparent abandonment of the area (Fig. 6.4).

She reported this issue on the App in hopes of getting support to work on a solution and bring this community space back to life for current and future generations. After being awarded the funds and executing the project, the beach and its neighboring park went from a derelict space covered in trash to a place that the community's proud of and uses to recreate and plan community activities. Motivated by this campaign, the community visited upstream polluters and has worked with them to stop polluting the river because it impacts where they play and surf. Interested readers can check out a video we posted about this on our website and YouTube channel, just look up Save The Waves Dirty Wave Challenge.

There are other great stories that have been supported by the App through the years that you can read in our App newsletters (at the bottom of our App webpage: savethewaves.org/app) and we hope that there will continue to be many more. From our experience with Hannah and other local partners, we're developing a program to support local leaders in their path to protect their surf ecosystems, starting from App training, outreach and coalition building to long-term planning, stakeholder engagement, and fundraising. The goal is that the App can be an "on-ramp" to surf conservation for surfers and coastal stewards around the world and that Save The Waves can cultivate and support these leaders. On its own, the App is only a tool, a technological asset, that will not protect any place by itself. However, when leveraged by the right people in the right places, it can be a truly powerful and effective way for communities to safeguard their valuable stretches of coastline.

The Growing Potential of AI

Looking ahead, there will continue to be highly impactful uses for technology and surf ecosystem conservation. We are already starting to see some groundbreaking work being done with Artificial Intelligence (AI). For better or worse, I do think this is the direction things are heading—where big money is moving and infrastructure is growing and therefore where people's time is going. I may not love the reality of the power demands of AI, or some of the ethical and political concerns around it, but I do not have control over that. Therefore, I am just trying to understand the applications for surf ecosystem conservation and use them to drive positive change.

A clear example of AI applied to surfing is Surfline Coastal Intelligence (previously called Surfzone AI). They are leveraging hundreds of their cameras with live streams from most major breaks around the world, including breaks that 10 or 15 years ago were considered "secret", to understand use patterns (i.e. number of surfers in the water and beach), surf conditions (how the wave changes in length, steepness, or size based on swell and tide), and improve the forecasting services they provide their users. As of the time of writing, this information has been leveraged by lifeguard associations, US State Parks, universities, and a few pioneering surf monitoring programs, and I'm certain that the list will continue to grow. This type of data opens the ability to predict how certain conditions might impact surf breaks' use and demonstrate the changes in surf break carrying capacity based on weather events, seasonal changes, or climate change–specifically large or unseasonable events like a big flood or a hurricane.

This application of AI is particularly helpful when considered in the context of climate finance markets, specifically around parametric insurance, which looks to create insurance policies that trigger a payout to beneficiaries of a natural resource (surf breaks, coral reefs, fisheries) based on parametrically-measurable environmental conditions (too much rain, very strong winds, hurricanes). In this case, imagine that a turbo-charged storm hits a community that depends on surf tourism during peak wave season. This leads to fewer visitors (in most cases) and trip cancellations, affecting the primary source of income for many people in this community. A parametric insurance product quickly provides these communities with funds to recover from the storm while surf tourists return to the area, paying out those whose business activities were disrupted by the extreme weather event. To set this up, historical data evidencing the linkages between storm events and number of surfers in lineups can corroborate the type of event that causes the most economic damage, and support fine-tuning of the product. Surfline Coastal Intelligence provides this and more which, while not a part of present-day insurance companies' list of approved datasets, could become a key data source to support surf ecosystem conservation based on wave quality and visitation across a variety of communities.

Visitor and usage information is also critical for economic valuations of surf ecosystems, also known as Surfonomics, where my interest for surf research began (savethewaves.org/surfonomics). In most global surfonomics studies, a combination of surveys and surfer counts of some type are utilized to estimate the direct expenditure from surf tourism in a surf community. While there are other methods that quantify the relative contribution of surfing to a coastal economy, this tends to be the most popular. It relies on visitation estimates from a range of sources, often dependent on data availability or budget to conduct in-person surveys, and tends to be a general estimate. With Surfline Coastal Intelligence, surf break usage can be estimated to an accuracy we've never had before, plus it can be done for various breaks at the same time. That's the exciting part about this use of AI–we can get actual information on how many surfers use surf breaks in various conditions and harness it to approximate the value and impacts on virtually any surf break around the world with more precision and fewer human resources than ever. This could help us understand the dynamic nature of the surf economy, surf ecosystems, and surf communities in general with increasing degrees of complexity as AI models and data processing power continue to improve and grow.

A great example where this application of AI is being leveraged is Las Flores in El Salvador, part of the newest Oriente Salvaje World Surfing Reserve dedicated in 2024, where Save The Waves is piloting a parametric insurance product—the first ever for a surf ecosystem. For example, as climate change exacerbates extreme weather events, torrential rains become more intense and frequent, leading to larger flooding events which reduce surfer visitation to Las Flores and interrupt the local economy that is dependent on surfing. Using Surfline Coastal Intelligence, we can look back at 10 years of historic use patterns (as long as the camera has been there) and quantify the reduction in number of surfers due to extreme weather events, and approximate the revenue lost due to any particular event through a combination of surfonomics data and this usage data. A threshold for extreme events can then be created based on the amount of precipitation and/or wind that leads to a disruption in the local surf economy. When this threshold is exceeded, an insurance policy payout is triggered, which goes to support the local community stakeholders that rely on surfing for their livelihood. While this is a first-of-its-kind pilot project, AI and surf camera data point to a future where we can better understand how changing conditions in the Anthropocene (like sea level rise, coastal erosion and more extreme weather events) impact surfing around the world. This will optimize efforts to address coastal issues and leverage financial mechanisms to account for increased climate risk across all dimensions of surf ecosystems (geophysical, social, biological), characterized in detail thanks to AI-supported technological advances.

What Does the Future Hold?

Technology is advancing extremely quickly. In just a few years the power of AI, driven by massive financial investments in processing power from technology companies, has led to a paradigm shift in the way people do their work and live their lives. I don't know where this is all going, but I expect more surf-related AI services to come to life, just like Surfline Coastal Intelligence but in a variety of sub-sectors related to surfing. Surf travel, surf parks, surf coaching, competitive surfing, surfboard manufacturing, among others, are some of many components of the surf industry where AI has direct, relatively obvious applications that could enable innovation and efficiency, and which will become cost-efficient soon.

Think of AI agents, autonomous systems powered by AI for a variety of tasks, as a (very smart) puppy learning from your feedback—what time do you usually go surfing? When do you take vacation? What kind of surf trips do you go on? You can also think of it as the most modern calculator money can buy, processing equations and code at speeds increasing every day, saving memory and adding capacity with each new model iteration, which has massive implications for surf and weather forecasting, creating more perfect and dynamic wave pools, and helping adequately manage coastlines around the world. Or you can imagine it as a second set of eyes, or perhaps more like bionic glasses, by allowing computer vision models to analyze surfers' movements, or surfboard construction, to a level of detail never seen before—potentially pushing the sport and surfboard design to even higher levels of performance and innovation. These are quite viable examples that may or may not already be going on at the time of writing, but that will surely become reality in the next few years. At that point, the latest technology advances, after the boom in AI investment, will be much further ahead and will enable even more applications to surfing and all of us who enjoy riding waves.

Will it come at the cost of people's jobs or traditional craft-making? Perhaps in some cases. While not everyone will upskill or become tech-savvy, I do believe that some efficiencies gained through technology will increase output and benefit the surf industry and its derivatives. However, it's anyone's guess who will benefit from it, in what capacity, and if it'll worsen or improve the environmental impact of surfing and the growing inequalities around the world. The one thing we can all count on is that things will change, perhaps very rapidly or perhaps not as fast as we all think, but hopefully they change for the better. It's up to us to do our share to ensure that as more technology is developed and made available to the masses, we are able to leverage it to better steward our environment, our waves, and our communities. Technology enables humanity, not the other way around, and it's important to keep that in mind in times of rapid technological advancement.

Perhaps if we put our rosy-colored glasses on, there's a future where we have a deep understanding of surf ecosystems and their multiple dimensions—a "digital twin" of a surf ecosystem. This means that we're able to model interactions between geophysical, social, and biological components of surf ecosystems all around the world to a degree of accuracy that mimics the real thing. Whether it's a reef break on a remote island of Indonesia, a sand-bottomed point break in Australia, or a thumping beach break in Central America, a "digital twin" would model the unique conditions that enable the surf ecosystem, and all the people and organisms that rely on it and make it unique. Imagine capturing that level of detail and complex webs of interactions and using them to effectively manage these places. Climate-driven issues, pollution, overcrowding, access, loss of critical habitat and species, and an equitable distribution of the benefits derived from surfing could all be effectively modeled and managed from the surf ecosystem approach, describing in detail to decision-makers the tradeoffs between coastal management approaches in the face of all of these threats and benefits. And the only way we'll get there is by leveraging the computational power behind AI, building artificial neural networks (like the ones in our brain) that can accurately replicate the web of complex dynamics in these ecosystems. It'll also take a lot of fieldwork to validate the models and ensure they match reality, which I am happy to volunteer for.

I don't know, think about it. Can we leverage technology to take care of surf ecosystems forever?

References

Arroyo, M., Levine, A., & Espejel, I. (2019). A transdisciplinary framework proposal for surf break conservation and management: Bahia de Todos Santos World Surfing Reserve. *Ocean and Coastal Managment, 168*, 197–211.

Arroyo, M., Seingier, G., Walter, G. P., & Gallegos, D. S. (2023). Surf conservation index: Mapping priority zones for surf protected areas in Mexico. *Shore & Beach, 91*(1), 49.

Sancho-Gallegos, D., & Arroyo Rodriguez, M. (2021). *Surf Conservation Index: Costa Rica.* Santa Cruz: Save the Waves Coalition. Retrieved from https://www.savethewaves.org/wp-content/uploads/2022/04/SurfConservationIndex_CostaRica_SaveTheWaves_2021-.pdf

Chapter 7
Unprovoked Shark Bites on Surfers: Trends in Occurrence and the Mitigation Options

Daryl McPhee

Introduction

Unprovoked shark bite over 30 years has been recorded from 56 countries and territories, the majority (84%) have occurred in six: the United States, South Africa, Australia, Brazil, the Bahamas and Reunion Island (Chapman & McPhee, 2016; McPhee, 2014). Unprovoked shark bite is a low probability high consequence event whose vivid nature skews risk perceptions (Sunstein, 2002). The threat of unprovoked shark bite is deeply ingrained in surf culture across many countries. Beach-goers, including surfers, often harbour a heightened fear of unprovoked shark bites, frequently overestimating the actual risk involved (Crossley et al., 2014; Lucrezi & Gennari, 2021). In terms of fatal unprovoked shark bites, the public overestimates their occurrence by a factor of four (Crossley et al., 2014). While the frequency of unprovoked shark bite has been increasing it is still low in absolute terms (two to three bites per year) (Huveneers et al., 2024; McPhee, 2014). The reasons for this increase are complex. While a greater number of water users over time contribute to this trend, it does not explain it entirely with several natural and anthropogenic factors contributing including climate change, increasing coastal prey (e.g. humpback whales) and coastal development (Amin, 2012; Chapman & McPhee, 2016; Ryan et al., 2019).

There has been a shift towards more unprovoked shark bites on surfers and less on swimmers. For example, in New South Wales (NSW, Australia) since 1900, bites shifted from being predominantly on swimmers to 79% on surfers by the 1980s (Huveneers et al., 2024). Such a shift reflects, at least in part, the increase in popularity of surfing. Three species of sharks are responsible for most serious unprovoked bites and fatalities—white sharks (*Carcharodon carcharias*), tiger sharks (*Galeocerdo cuvier*) and bull sharks (*Carcharhinus leucas*) (McCourt, 2023; McPhee, 2014; Riley

D. McPhee (✉)
Faculty of Society and Design, Bond University, Robina, QLD, Australia
e-mail: dmcphee@bond.edu.au

L. Mach (ed.), *The Future of Surfing in the Anthropocene*,
https://doi.org/10.1007/978-3-032-09377-6_7

et al., 2022). In NSW, almost half of the recorded bites on surfers are attributed to white sharks (McPhee et al., 2022).

Unprovoked shark bite not only impacts the victims, but their families, first responders, eyewitnesses and the local surfing community (Taylor et al., 2019). While the probability of an unprovoked shark bite remains low, the vivid nature of a shark bite ensures a high degree of media reporting and public concern (Losen, 2023; Neff, 2012). This is heightened if a series of bites occurs in a region over a short period of time (Barnett et al., 2022; McPhee, 2014). Series of bites on surfers over a relatively short period of time have occurred in Australia (northern NSW and Western Australia (WA)), Brazil and Reunion. Public discourse in response is often highly polarised with calls for shark culls from some which are opposed by many others (Fraser-Baxter & Medvecky, 2018). As an example, two of the world's top surfers Kelly Slater and Jeremy Flores called for a shark cull at Reunion following a series of fatal bites and this call met with criticism.[1]

In this chapter, recent trends in the number and severity of unprovoked shark bites on surfers are provided. The role technology plays in shark bite mitigation are described and future avenues of technological advancement to mitigate risk are discussed.

Trends in Unprovoked Shark Bites on Surfers

Table 7.1 documents trends in the number of surfers bitten by sharks in countries and jurisdictions within countries where surfing is known to be popular. This information was extracted from the Global Shark Attack File (*Global Shark Attack File home page*). Consistent with Lentz et al. (2010), each incident was given an index of severity from one (minor injuries) to five (fatal). The average index of severity was then calculated for each country/state and time period and appears in parentheses in Table 7.1.

Overall, unprovoked shark bites on surfers are rare but trends are evident. Trends in unprovoked shark bite on surfers differs geographically. There has been a doubling of incidents in Australia between the decades 2000–09 and 2010–19. This overall increase is driven by an increase in bites in the states of NSW and WA with bites in other states remaining relatively low over the period examined. There is no clear trend in the severity of bites in Australia. In the US, the number of bites is dominated by those that occur in Florida. Bites on surfers in Florida are less severe than other locations and this is potentially due to them being from smaller whaler species (e.g. black tip whalers, *Carcharhinus limbatus*) and no fatalities have been recorded (McPhee, 2014). Bites in Hawaii have more than doubled between the decades 2000–09 and 2010–19, but the severity of bites has remained similar. Bites in California were more prevalent during 2000–09 but increased in severity in 2010–19. Bites in

[1] Surf champion Kelly Slater calls for daily shark cull on Réunion | Réunion | The Guardian (Accessed 25/1/25).

Table 7.1 Decadal and geographic trends in the number and severity of shark bites on surfers in countries (and states within Australia and the US)

Time period	1980–89	1990–1999	2000–2009	2010–2019
Location				
Australia	6 (2.83), 2 fatal	9 (1.78), 1 fatal	26 (2.50), 3 fatal	53 (2.33), 7 fatal
NSW	*2 (3.50), 1 fatal*	*1 (2.00)*	*15 (2.29), 1 fatal*	*30 (2.13), 2 fatal*
WA	*1 (2.00)*	–	*1 (5.00), 1 fatal*	*12 (3.60), 5 fatal*
Qld	–	*3 (2.67), 1 fatal*	*2 (1.00)*	*2 (1.00)*
SA	*2 (3.50), 1 fatal*	*4 (1.20)*	*5 (3.20), 1 fatal*	*3 (2.70)*
Other Aus states	*1 (1.00)*	*1 (1.00)*	*4 (2.50)*	*6 (1.00)*
United States	56 (1.33), 1 fatal	80 (1.49), 1 fatal	151 (1.42), 2 fatal	130 (1.67), 1 fatal
Florida	*46 (1.22)*	*60 (1.35)*	*122 (1.26)*	*95 (1.51)*
Hawaii	–	*7 (2.16), 1 fatal*	*7 (2.43), 1 fatal*	*18 (2.22)*
California	*4 (2.00), 1 fatal*	*6 (2.29)*	*15 (1.80)*	*7 (2.71), 1 fatal*
Oregon	*2 (1.50)*	*3 (1.67)*	*3 (2.00)*	*1 (3.00)*
Other US states	*4 (1.00)*	*4 (1.00)*	*4 (2.50), 1 fatal*	*9 (1.38)*
South Africa	10 (2.78), 1 fatal	16 (2.56), 3 fatal	15 (2.40), 2 fatal	17 (2.41), 2 fatal
Reunion	1 (1.00)	5 (4.75), 3 fatal	3 (3.33), 1 fatal	10 (4.00), 2 fatal
Brazil	–	23 (3.72), 5 fatal	7 (3.80), 1 fatal	3 (2.00)

South Africa have remained relatively consistent in number and severity across the four decades considered. Unprovoked shark bites at Reunion are higher in severity than other locations and peaked in number between 2010 and 19. In Brazil, there is a clear peak in the number of bites in the 1990–99 period and these bites on average were severe with five fatalities recorded during that period. Bites in Brazil mostly occur at a single location (Recife, Pernambuco) (Hazin & Burgess, 2008).

Conceptualising Shark Bite Mitigation for Surfers

Shark bite mitigation programs are a particularly challenging and problematic case for politicians and managers where conflicting views within a community are heightened by the sensitive topic of human life versus species protection (Cullen-Knox et al., 2017). The public views shark bite mitigation as a combination of personal accountability and government responsibility (Lucrezi et al., 2021), although some form of government response is generally required (Simmons et al., 2021). Effective mitigation requires approaches that reduce the risk of a bite occurring, but which also provides peace of mind for the surfer.

Overall, there is no single universal approach that beach authorities or individuals can use to mitigate unprovoked shark bite. While no approach will ever be 100% effective, a reduction in the risk of shark bite is achievable. Approaches to mitigation can involve a combination of technology, education and a level of personal responsibility from a surfer themselves. Huveneers et al. (2024) describes a hierarchical approach to mitigating unprovoked shark bite which consists of three components:

- *Perimeter: reduce spatial overlap between water users and sharks*—The aim is to reduce or eliminate the probability of humans and sharks coming into contact by reducing their spatiotemporal overlap. This can be achieved by physically separating humans and sharks (e.g., physical enclosures), reducing local shark abundance (e.g., beach meshing, culls), using early warning systems (e.g., listening stations, SMART drumlines, aerial surveillance, land-based spotting programs), or changing human behaviour (avoid times and locations with a high probability of relatively higher shark abundance).
- *Proximity: reduce likelihood of shark bite when overlap cannot be avoided*—The aim is to deter an approaching shark from biting. Several types of shark deterrent exist and are commercially available, including electric, magnetic, chemical, and visual deterrents that are predicated on overwhelming a shark's sensory organs or disorienting the animal (Hart & Collin, 2015). For example, personal electric deterrents aim to produce an electric field strong enough to overstimulate a shark's ampullae of Lorenzini (specialized sensory organs found in sharks), and thus, change the trajectory of the animal through causing discomfort and/or confusion.
- *Prevent bleeding: reduce injuries from shark bites when sharks cannot be deterred*—The aim is to reduce injuries and recovery times when a bite occurs. Blood loss is the leading cause of fatality from shark bites (Ballas et al., 2017; Lentz et al., 2010; Woolgar et al., 2001), and so mitigation measures and emergency responses that minimise blood loss can reduce the probability of dying from a shark bite.

Perimeter Measures for Protecting Surfers

In the Australian states of Queensland and NSW and in KwaZulu-Natal (South Africa) nets and drumlines have been historically used to attempt to reduce the local abundance of sharks at popular beaches by lethal methods (Dudley, 1997). These approaches have become increasingly controversial due to changing community attitudes (including from surfers) regarding the impacts of the program on target sharks and other marine species (e.g. whales and dolphins) and no longer has widespread support (Atkins et al., 2023; Gibbs & Warren, 2015; Gibbs et al., 2020; Meeuwig & Ferreira, 2014; Sheridan et al., 2021). This change in attitude is a result of an understanding of the conservation status of several species, the role of some shark species as apex predators, and recognition of the intrinsic value and ecosystem benefits of sharks (Simpfendorfer et al., 2011; O'Connell & de Jonge, 2014; Gibbs & Warren, 2015; Pepin-Neff & Wynter, 2019; Couper & Walters, 2021; Sheridan et al., 2021). It is expected that there will be a clear continuing transition away from these lethal methods where they occur and little scope to expand them into other jurisdictions.

Many jurisdictions have invested heavily in developing and trialling non-lethal approaches to mitigate unprovoked shark bite with significant focus on technology that can be applied locally to reduce the potential overlap between large shark species and water users (McPhee et al., 2022). The efficacy of mitigation measures is frequently location specific.

As a technological change to the traditional baited drumline, the Shark-Management-Alert-in-Real-Time (SMART) drumline was invented at Reunion as a mitigation method that can maximise survival rate of captured animals (Guyomard et al., 2019; McPhee et al., 2021). SMART drumlines include a GPS satellite buoy and a 'Catch-A-Live'™ system alerting contractors in near real-time to when an animal is caught. The alert is initiated when the bait is taken by an animal and the magnetic trigger is activated. Sharks captured on SMART drumlines are relocated alive to offshore locations. The approach constitutes a meaningful compromise between the need to target potentially dangerous species near popular beaches and the requirement to safely release, and possibly tag captured animals (Guyomard et al., 2019). SMART drumlines have been found to be effective at catching white sharks in NSW (Tate et al., 2021), however they were ineffective in WA for this species (Taylor et al., 2022). The specific reasons for this difference in efficacy are unclear but it highlights the need to tailor mitigation approaches to specific local conditions.

Purpose designed physical barriers can be used at a location to separate sharks from water users and have been successfully used at some WA beaches and Fish Hoek Bay (Cape Town, South Africa) (McPhee et al., 2021). The barriers enclose a beach area, thus preventing sharks from entering an area of water where people are present. However, physical barriers are difficult to successfully deploy and maintain at most surf breaks as wave action can damage or completely dislodge them. Further, there can be concerns from surfers because depending on local factors, a physical barrier may become a collision or entanglement risk.

Unmanned Aerial Vehicles (UAVs of "drones") are increasingly being used to detect sharks along popular beaches with detections subsequently passed on in close to real time to inform water users of their presence (Butcher et al., 2021; McPhee et al., 2021; Stokes et al., 2020). The use of UAVs has largely replaced the use of fixed wing aircraft which is more expensive and less effective (McPhee et al., 2021). The use of tethered UAVs via helium filled balloons and/or kites may provide opportunity for longer-term surveillance at a specific location beyond what is possible for remotely piloted drones (Kiszka et al., 2018). Development of algorithms and the application of AI to automate accurate detection of potentially dangerous species are possible and may improve the accuracy of detections (Butcher et al., 2021; Colefax et al., 2019; Gorkin et al., 2020; Sharma et al., 2018), but surface chop and water clarity could still prove too complex to be reliable under the variable conditions usually experienced within one day at a beach. However, the use of UAVs with multispectral cameras shows promise as a tool for shark detection to address this limitation as this technology can be effective in a much wider variety of ambient conditions (Butcher et al., 2021; Chirayath & Earle, 2016). Multispectral cameras collect visual information outside the range of the human eye which is then processed to provide a visual representation. While there is proof of concept of the approach, there is yet to be a comprehensive field trial of this technology for the purpose of shark bite mitigation (Butcher et al., 2021).

The acoustic tagging and subsequent detection of tagged sharks is a method that provides important information on the movement patterns of sharks that can be used in the long-term to gain an understanding of when and where relevant shark species are likely to be, including whether they are in the vicinity of water users (e.g. Niella et al., 2020; Smoothey et al., 2019). Tagged sharks are detected at acoustic receivers (listening stations) and these can be deployed individually or as an array along a stretch of coastline. As an example of the importance of information that can be collected for informing risk to surfers, Smoothey et al. (2023) identified that in NSW the occurrence of large bull sharks in nearshore waters was greatest from midday to 04:00 pm, when water temperatures were higher than 20 °C, after > 45 mm of rain and when swell heights were between 1.8 and 2.8 m. This provides a very specific set of parameters for when bull sharks are most likely to be present in the nearshore area of that region and thus when risk of an unprovoked shark bite occurring from that species is elevated. This demonstrates how the marriage of technological advances with scientific evaluation can provide useful general safety information for lifesaving personnel and governments to pass along to surfers and help to mitigate risk.

When tagged sharks are detected at acoustic receivers (listening stations), detections in real time (or close to it) can be communicated to the public via social media, dedicated websites, through surf life savers and text messages to subscribers (McPhee et al., 2021). The direct efficacy of acoustic receivers as a tool for mitigating unprovoked shark bite requires ongoing efforts to tag large sharks for subsequent detection. Although not yet currently well assessed, there is a risk of surfers becoming either apathetic to very frequent shark alerts or experiencing heightened fear.

The use of sonar technologies can circumvent the need to tag sharks to detect them and has been considered for use in shark bite mitigation. Sonar technologies, in particular detection algorithms, have advanced substantially in recent years but have still fallen short of being an effective approach for shark bite mitigation (McPhee et al., 2021; Parsons et al., 2014). Acoustic propagation in the surf zone is a challenging technical problem due to its shallow nature and the presence of air bubbles and sand in the water column (Farmer et al., 2001; Vagle et al., 2005). As sonar units are expensive, the number of units potentially needed to provide detection at a beach is an important consideration in terms of the cost effectiveness of the technology if applied in practice.

When considering the spatial overlap between sharks and surfers at a large geographic scale, climate induced changes are altering the distribution of relevant shark species. For example, the geographic range of bull and tiger sharks on the Australian east coast are extending and/or forecast to extend further south (poleward) which will potentially increase the geographic range of overlap between surfers and these shark species (Smoothey et al., 2023; Niella et al., 2020).

Proximity Measures for Mitigation

Personal protection devices (personal shark deterrents) are a potentially effective and important part of the toolbox of shark-bite mitigation measures. They are one approach an individual surfer can use to take additional personal responsibility for their safety. Personal shark deterrents can be based on electric, magnetic, visual and chemical approaches. Entrepreneurial activity associated with the development of personal shark deterrents is high in countries such as Australia. Modelling has predicted that across Australia, > 1000 people could potentially avoid being bitten by 2066 if all water-users wore effective personal shark deterrents (Bradshaw et al., 2021).

Personal electric shark deterrents that either clip on to the surfboard or are embedded in the surfboard have been shown in several studies to reduce the risk of a bite occurring by $\sim 60\%$ (Blount et al., 2021; Huveneers et al., 2013, 2018; Smit & Peddemors, 2003; Thiele et al., 2020). Electric shark deterrents rely on overstimulating the ampullae of Lorenzeni which are specialised sense organs that sharks (and rays) have for detecting electric currents that are lacking in other marine animals. This specificity means that electric shark deterrents do not impact other marine animals. While most of the testing of personal electric deterrents suitable for surfers has focussed on white sharks (e.g. Blount et al., 2021; Huveneers et al., 2013, 2018; Kempster et al., 2016), more recent work has also included tiger and bull sharks. Clarke et al. (2024) found that the Ocean Guardian Freedom Surf+[2] designed specifically for clipping on to a surfboard reduced the risk of a bite occurring by 54% for all three relevant species. The Rpela v2 device is a battery-powered device embedded within the surfboard and field testing found a 66% reduction in the probability of a white shark bite occurring with the device active (Blount et al., 2021).

Sharks can detect magnetic fields, although there is uncertainty regarding the sensory receptors and pathways involved in detecting these fields and the exact nature of perceived stimuli (Anderson et al., 2017). Some shark species use information on magnetic fields for migration (Keller et al., 2021). Magnets have been used to attempt to reduce shark bycatch in commercial fisheries with some success (Richards et al., 2018; Rigg et al., 2009; Robbins et al., 2011). However, examples have been recorded where there is no significant reduction in shark bycatch from the use of magnets (Polpetta et al., 2021) and an example of where shark bycatch increased (Porsmoguer et al., 2015).

The challenge with using magnetic products as a personal shark deterrent is that a small magnet that is practical to wear or place on a board will only produce a magnetic field over a very small area. Huveneers et al. (2018) identified that magnets of the size and type used as personal shark deterrents can only be detected by a shark when it is approximately 30 cm away from it and this clearly reduces their effectiveness.

[2] This product is no longer commercially available at the time of writing.

Sharks either have poor colour vision or are completely colour blind but respond to high visual contrast (Hart & Collin, 2015). There is no one "colour" that a shark will respond to or be deterred by consistently as the contrast as seen by a shark will vary based on overhead conditions (e.g. time of day, cloud cover). However, a colour such as yellow which has high contrast against most backgrounds would be expected to be more visible to some shark species more often (Hart & Collin, 2015).

Counterillumination that reduces the silhouette of a surfer and surfboard from below is a potential avenue for the development of a novel and effective visual shark deterrent. Using an array of LED lights to change the pattern of counterillumination, Ryan et al. (2024) found that the probability of a bite on a seal decoy by a white shark could be significantly reduced. Further work is required to test the effectiveness of this approach in different contexts. If there is broader effectiveness, the development of a commercially available deterrent based on using lighting to change the pattern of counterillumination is a future development step for this potential approach.

The use of chemical shark deterrents has a long history, first being developed in 1942 in response to the fear of sharks held by US servicemen during World War 2 (Hart & Collin, 2015; McPhee et al., 2021). This early approach was not effective (McPhee et al., 2021). A shark deterrent surfboard wax consisting of various oils (e.g. eucalyptus oil) and spices (e.g. chilli, cayenne pepper and cloves) is commercially available but independent testing found it to it be ineffective (Huveneers et al., 2018). None of the ingredients used in this wax is by itself an established shark deterrent compound and so the product is unlikely to dissuade a shark from biting a surfer (Huveneers et al., 2018).

Preventing Bleeding for Mitigation

Blood loss is what leads to a shark bite becoming fatal in many instances. Appropriate and timely first aid that can reduce bleeding following a bite remains an important tool for minimising the chance of a serious bite becoming fatal. Advancements in fabric technologies provide additional options for reducing blood loss. Wetsuit materials that can largely withstand a shark bite have been developed (Whitmarsh et al., 2019; Fielder & Verstegen, 2020). These wetsuits do not change the probability of a bite occurring but can reduce the consequence. While crushing injuries, skin punctures, and lacerations are still possible, these protective wetsuits can help reduce blood loss, providing more time for medical aid to be given to the victim. The advantage of a wetsuit for surfers is that they will already use wetsuits in many circumstances so they are not required to have an additional piece of gear such as a personal shark deterrent, but rather they can substitute an existing piece of gear.

Discussion

With the continued growth in the popularity of surfing, it is expected that the number of unprovoked shark bites will continue to rise, but the overall risk to an individual at any one time of a bite occurring will continue to be low. No single approach to mitigating the risk of unprovoked shark bite on surfers exists and the risk cannot be eliminated. A combination of non-lethal approaches is required to reduce risk, and the efficacy and practicality of approaches will vary based on local biophysical conditions and local surf culture. Technological developments are expected to continue to contribute to the efficacy and cost-effectiveness of unprovoked shark bite mitigation.

There are a range of approaches that have been developed and implemented over the past decade (e.g. UAVs and telemetry) which have a positive role in shark bite mitigation. A fundamental issue with quantifying the effectiveness of any method is the challenge of reliably assessing the direct reduction in shark bite risk due to the rarity of incidents. However, technologies are likely to continue to advance in key properties (e.g. battery life in UAVs) which will improve their efficacy, and they are likely to become more cost-effective and their use more widespread. New approaches will continue to be identified through research activity and at least some of these should have the potential to be commercialised and add to the options currently available.

A challenge for the use of mitigation technologies and approaches is their acceptance and uptake by surfers. As an example, in NSW surfers indicated lower confidence in the utility of UAVs for shark bite mitigation compared to other beach users (Stokes et al., 2020). In the case of personal electric deterrents despite the studies that have shown a statistically significant reduction in the probability of a shark bite occurring when using them, there can be resistance in the surfing community to their use. Simmons et al. (2021) identified that personal shark deterrents were considered expensive and cumbersome and not many people trusted them. There is a belief among some surfers that an electric deterrent device attracts sharks, but such a view is inconsistent with the evidence available (Clarke et al., 2024). Engagement with surfers at the local level can assist with understanding barriers and assist uptake of approaches that demonstrate technical merit.

A diversity of personal shark deterrents are available for surfers to purchase, many of which have not been independently tested for effectiveness. This creates an additional challenge for surfers wanting to purchase a personal shark deterrent in terms of determining which are effective. Using a personal shark deterrent that is ineffective may create a false sense of security for a surfer. For other safety devices such as car seatbelts, life jackets and bike helmets, a minimum mandated standard of effectiveness needs to be met before they can be sold as a safety device. Currently lacking are relevant technical standards for manufacturers to meet to be able to reliably claim that a personal shark deterrent is effective and to provide surfers (and other water users) with confidence that the deterrents will reduce the risk of a bite occurring.

Conclusion

While the probability of an unprovoked shark bite occurring on a surfer is low, it is increasing as the popularity of surfing continues to increase. Climate induced changes to the marine environment will continue to alter the geographic distribution of sharks and the risk of unprovoked shark bite on surfers would be expected to change geographically in response including increased risk in some areas such south-eastern Australia.

There is (and will continue to be) no single universal approach for mitigating the risk of an unprovoked shark bite, but a suite of approaches can be used and tailored to local beach environments and individual surfers to reduce the risk of unprovoked shark bite. A missing requirement for providing surfers confidence in the effectiveness of personal shark deterrent approaches is the development and application of relevant safety standards and independent testing of any personal shark deterrent against a standard. A standard is critical so that effective personal shark deterrents can be distinguished from ineffective ones.

Drone technologies are advancing with factors such as flight times increasing rapidly which will increase the utility of their use for mitigating unprovoked shark bite on surfers. While current drone technology is focussed on assessment in the visual spectrum, new camera technologies that operate outside the visual spectrum in combination with AI technologies will see further improvement in the efficacy of drone technology.

In terms of personal shark deterrents, electric deterrents have been demonstrated to be effective and further refinements to these types of deterrents to improve effectiveness will likely continue. Visual deterrents based on counterillumination to reduce the silhouette of a surfer and surfboard are a promising avenue of future research. Personal shark deterrents based on magnets and chemicals are limited in their effectiveness (or ineffective) and this is likely to be the case moving forward.

Regardless of the low probability of an unprovoked shark bite occurring on a surfer and the effectiveness of mitigation, sharks will remain vividly embedded in surf culture in many countries.

References

Amin, R., Ritter, E., & Kennedy, P. (2012). A geospatial analysis of shark attack rates for the east coast of Florida: 1994–2009. *Marine and Freshwater Behaviour and Physiology, 45*(3), 185–198.

Anderson, J. M., Clegg, T. M., Véras, L. V., & Holland, K. N. (2017). Insight into shark magnetic field perception from empirical observations. *Scientific Reports, 7*(1), 11042.

Atkins, S., Mann-Lang, J., Cliff, G., Pillay, N., & Cantor, M. (2023). Stakeholder perceptions reveal obstacles and opportunities to change lethal methods of protecting bathers from sharks. *Marine Policy, 155*, 105762.

Ballas, R., Saetta, G., Peuchot, C., Elkienbaum, P., & Poinsot, E. (2017). Clinical features of 27 shark attack cases on La Réunion Island. *Journal of Trauma and Acute Care Surgery, 82*(5), 952–955.

Barnett, A., Fitzpatrick, R., Bradley, M., Miller, I., Sheaves, M., Chin, A.,... & Abrantes, K. (2022). Scientific response to a cluster of shark bites. *People and Nature, 4*(4), 963–982.

Blount, C., Pygas, D., Smith, M. P. L., McPhee, D. P., Bignell, C., & Ramsey, O. (2021). Effectiveness against white sharks of the rpela personal shark deterrent device designed for surfers. *Journal of Marine Science and Technology (Taiwan), 29*(4), 582–591.

Bradshaw, C. J., Meagher, P., Thiele, M. J., Harcourt, R. G., & Huveneers, C. (2021). Predicting potential future reduction in shark bites on people. *Royal Society Open Science, 8*(3), 201197.

Butcher, P. A., Colefax, A. P., Gorkin III, R. A., Kajiura, S. M., López, N. A., Mourier, J.,... & Raoult, V. (2021). The drone revolution of shark science: A review. *Drones, 5*(1), 8.

Chapman, B. K., & McPhee, D. (2016). Global shark attack hotspots: Identifying underlying factors behind increased unprovoked shark bite incidence. *Ocean & Coastal Management, 133*, 72–84.

Chirayath, V., & Earle, S. A. (2016). Drones that see through waves–preliminary results from airborne fluid lensing for centimetre-scale aquatic conservation. *Aquatic Conservation: Marine and Freshwater Ecosystems, 26*, 237–250.

Clarke, T. M., Barnett, A., Fitzpatrick, R., Ryan, L. A., Hart, N. S., Gauthier, A. R.,... & Huveneers, C. (2024). Personal electric deterrents can reduce shark bites from the three species responsible for the most fatal interactions. *Scientific Reports, 14*(1), 16307.

Colefax, A. P., Butcher, P. A., Pagendam, D. E., & Kelaher, B. P. (2019). Reliability of marine faunal detections in drone-based monitoring. *Ocean & Coastal Management, 174*, 108–115.

Couper, A., & Walters, R. (2021). The great white bite: A critique of the Western Australian government's shark hazard mitigation drum line program. *Journal of Sociology, 57*(4), 842–859.

Crossley, R., Collins, C. M., Sutton, S. G., & Huveneers, C. (2014). Public perception and understanding of shark attack mitigation measures in Australia. *Human Dimensions of Wildlife, 19*(2), 154–165.

Cullen-Knox, C., Haward, M., Jabour, J., Ogier, E., & Tracey, S. R. (2017). The social licence to operate and its role in marine governance: Insights from Australia. *Marine Policy, 79*, 70–77.

Dudley, S. F. (1997). A comparison of the shark control programs of New South Wales and Queensland (Australia) and KwaZulu-Natal (South Africa). *Ocean & Coastal Management, 34*(1), 1–27.

Farmer, D. M., Deane, G. B., & Vagle, S. (2001). The influence of bubble clouds on acoustic propagation in the surf zone. *IEEE Journal of Oceanic Engineering, 26*(1), 113–124.

Fiedler, T., & Verstegen, T. (2020). Fibre-reinforced composite for protection against shark bites. *Materials, 13*(22), 5065.

Fraser-Baxter, S., & Medvecky, F. (2018). Evaluating the media's reporting of public and political responses to human-shark interactions in NSW, Australia. *Marine Policy, 97*, 109–118.

Gibbs, L., Fetterplace, L., Rees, M., & Hanich, Q. (2020). Effects and effectiveness of lethal shark hazard management: The shark meshing (bather protection) program, NSW, Australia. *People and Nature, 2*(1), 189–203.

Gibbs, L., & Warren, A. (2015). Transforming shark hazard policy: Learning from ocean-users and shark encounter in Western Australia. *Marine Policy, 58*, 116–124.

Gorkin, R., III., Adams, K., Berryman, M. J., Aubin, S., Li, W., Davis, A. R., & Barthelemy, J. (2020). Sharkeye: Real-time autonomous personal shark alerting via aerial surveillance. *Drones, 4*(2), 18.

Guyomard, D., Perry, C., Tournoux, P. U., Cliff, G., Peddemors, V., & Jaquemet, S. (2019). An innovative fishing gear to enhance the release of non-target species in coastal shark-control programs: The SMART (shark management alert in real-time) drumline. *Fisheries Research, 216*, 6–17.

Hart, N. S., & Collin, S. P. (2015). Sharks senses and shark repellents. *Integrative Zoology, 10*(1), 38–64.

Hazin, F. H., Burgess, G. H., & Carvalho, F. C. (2008). A shark attack outbreak off Recife, Pernambuco, Brazil: 1992–2006. *Bulletin of Marine Science, 82*(2), 199–212.

Huveneers, C., Rogers, P. J., Semmens, J. M., Beckmann, C., Kock, A. A., Page, B., & Goldsworthy, S. D. (2013). Effects of an electric field on white sharks: In situ testing of an electric deterrent. *PLoS ONE, 8*(5), e62730.

Huveneers, C., Whitmarsh, S., Thiele, M., Meyer, L., Fox, A., & Bradshaw, C. J. (2018). Effectiveness of five personal shark-bite deterrents for surfers. *PeerJ, 6*, e5554.

Huveneers, C., Blount, C., Bradshaw, C. J., Butcher, P. A., Smith, M. P. L., Macbeth, W. G.,... & Green, M. (2024). Shifts in the incidence of shark bites and efficacy of beach-focussed mitigation in Australia. *Marine Pollution Bulletin, 198*, 115855.

Keller, B. A., Putman, N. F., Grubbs, R. D., Portnoy, D. S., & Murphy, T. P. (2021). Map-like use of Earth's magnetic field in sharks. *Current Biology, 31*(13), 2881–2886.

Kempster, R. M., Egeberg, C. A., Hart, N. S., Ryan, L., Chapuis, L., Kerr, C. C.,... & Collin, S. P. (2016). How close is too close? The effect of a non-lethal electric shark deterrent on white shark behaviour. *PLoS One, 11*(7), e0157717.

Kiszka, J. J., Heithaus, M. R., Carrier, J. C., & Simpfendorfer, C. A. (2018). Using aerial surveys to investigate the distribution, abundance, and behavior of sharks and rays. In: *Shark Research: Emerging Technologies and Applications for the Field and Laboratory*, 71–82.

Lentz, A. K., Burgess, G. H., Perrin, K., Brown, J. A., Mozingo, D. W., & Lottenberg, L. (2010). Mortality and management of 96 shark attacks and development of a shark bite severity scoring system. *The American Surgeon, 76*(1), 101–106.

Losen, B. (2023). Shark attack risk on Reunion Island: Emphasis on local media construction. *Marine Policy, 157*, 105851.

Lucrezi, S., & Gennari, E. (2021). Perceptions of shark hazard mitigation at beaches implementing lethal and nonlethal shark control programs. *Society & Animals, 30*(5–6), 646–667.

McCourt, J. A. (2023). Fatal shark attacks. *Pathology, 55*, S10.

McPhee, D. (2014). Unprovoked shark bites: Are they becoming more prevalent? *Coastal Management, 42*(5), 478–492.

McPhee, D. P., Blount, C., Smith, M. P. L., & Peddemors, V. M. (2021). A comparison of alternative systems to catch and kill for mitigating unprovoked shark bite on bathers or surfers at ocean beaches. *Ocean & Coastal Management, 201*, 105492.

McPhee, D., Blount, C., & MacBeth, W. (2022). *NSW shark management strategy and shark program review.* A report prepared for the NSW Department of Regional NSW.

Meeuwig, J. J., & Ferreira, L. C. (2014). Moving beyond lethal programs for shark hazard mitigation. *Animal Conservation, 17*(4), 297–298.

Neff, C. (2012). Australian beach safety and the politics of shark attacks. *Coastal Management, 40*(1), 88–106.

Niella, Y., Smoothey, A. F., Peddemors, V., & Harcourt, R. (2020). Predicting changes in distribution of a large coastal shark in the face of the strengthening East Australian Current. *Marine Ecology Progress Series, 642*, 163–177.

O'Connell, C. P., & de Jonge, V. N. (2014). Integrating the findings from this special issue and suggestions for future conservation efforts—A brief synopsis. *Ocean & Coastal Management, 97*, 58–60.

Parsons, M. J., Parnum, I. M., Allen, K., McCauley, R. D., & Erbe, C. (2014). Detection of sharks with the Gemini imaging sonar. *Acoustics Australia, 42*(3), 185–190.

Pepin-Neff, C., & Wynter, T. (2019). Save the sharks: Reevaluating and (re) valuing feared predators. *Human Dimensions of Wildlife, 24*(1), 87–94.

Polpetta, M., Piva, F., Gridelli, S., & Bargnesi, F. (2021). Behavioural responses in the sand tiger shark (*Carcharias taurus*) to permanent magnets and pulsed magnetic fields. *Marine Biology Research, 17*(1), 41–56.

Porsmoguer, S. B., Bănaru, D., Boudouresque, C. F., Dekeyser, I., & Almarcha, C. (2015). Hooks equipped with magnets can increase catches of blue shark (*Prionace glauca*) by longline fishery. *Fisheries Research, 172*, 345–351.

Richards, R. J., Raoult, V., Powter, D. M., & Gaston, T. F. (2018). Permanent magnets reduce bycatch of benthic sharks in an ocean trap fishery. *Fisheries Research, 208*, 16–21.

Rigg, D. P., Peverell, S. C., Hearndon, M., & Seymour, J. E. (2009). Do elasmobranch reactions to magnetic fields in water show promise for bycatch mitigation? *Marine and Freshwater Research, 60*(9), 942–948.

Riley, M., Meagher, P., Huveneers, C., Leto, J., Peddemors, V. M., Slip, D.,... & Bradshaw, C. J. (2022). The Australian shark-incident database for quantifying temporal and spatial patterns of shark-human conflict. *Scientific Data, 9*(1), 378.

Robbins, W. D., Peddemors, V. M., & Kennelly, S. J. (2011). Assessment of permanent magnets and electropositive metals to reduce the line-based capture of Galapagos sharks, *Carcharhinus galapagensis*. *Fisheries Research, 109*(1), 100–106.

Ryan, L. A., Lynch, S. K., Harcourt, R., Slip, D. J., Peddemors, V., Everett, J. D.,... & Hart, N. S. (2019). Environmental predictive models for shark attacks in Australian waters. *Marine Ecology Progress Series, 631*, 165–179.

Ryan, L. A., Gennari, E., Slip, D. J., Collin, S. P., Peddemors, V. M., Huveneers, C.,... & Hart, N. S. (2024). Counterillumination reduces bites by Great White Sharks. *Current Biology, 34*(24), 5789–5795.

Sharma, N., Scully-Power, P., & Blumenstein, M. (2018). Shark detection from aerial imagery using region-based CNN, a study. In *AI 2018: Advances in Artificial Intelligence: 31st Australasian Joint Conference, Wellington, New Zealand, December 11–14, 2018, Proceedings 31* (pp. 224–236). Springer International Publishing.

Sheridan, K., O'Riain, M. J., & Needham, M. D. (2021). Recreationist perceptions of lethal and non-lethal management of sharks in two of South Africa's marine areas. *Marine Policy, 132*, 104633.

Simmons, P., Mehmet, M., Curley, B., Ivory, N., Callaghan, K., Wolfenden, K., & Xie, G. (2021). A scenario study of the acceptability to ocean users of more and less invasive management after shark-human interactions. *Marine Policy, 129*, 104558.

Simpfendorfer, C. A., Heupel, M. R., White, W. T., & Dulvy, N. K. (2011). The importance of research and public opinion to conservation management of sharks and rays: A synthesis. *Marine and Freshwater Research, 62*(6), 518–527.

Smit, C. E., & Peddemors, V. (2003). Estimating the probability of a shark attack when using an electric repellent: Applications. *South African Statistical Journal, 37*(1), 59–78.

Smoothey, A. F., Lee, K. A., & Peddemors, V. M. (2019). Long-term patterns of abundance, residency and movements of bull sharks (*Carcharhinus leucas*) in Sydney Harbour, Australia. *Scientific Reports, 9*(1), 18864.

Smoothey, A. F., Niella, Y., Brand, C., Peddemors, V. M., & Butcher, P. A. (2023). Bull Shark (*Carcharhinus leucas*) occurrence along beaches of South-Eastern Australia: Understanding where, when and why. *Biology, 12*(9), 1189.

Stokes, D., Apps, K., Butcher, P. A., Weiler, B., Luke, H., & Colefax, A. P. (2020). Beach-user perceptions and attitudes towards drone surveillance as a shark-bite mitigation tool. *Marine Policy, 120*, 104127.

Sunstein, C. (2002). Probability neglect: Emotions, worst case scenarios, and law. *The Yale Law Journal, 112*, 61–107.

Tate, R. D., Kelaher, B. P., Brand, C. P., Cullis, B. R., Gallen, C. R., Smith, S. D., & Butcher, P. A. (2021). The effectiveness of Shark-Management-Alert-in-Real-Time (SMART) drumlines as a tool for catching white sharks, *Carcharodon carcharias*, off coastal New South Wales, Australia. *Fisheries Management and Ecology, 28*(5), 496–506.

Taylor, J., McLean, L., Korner, A., & Glozier, N. (2019). Direct and indirect psychological impacts of shark-bite events. *Australian & New Zealand Journal of Psychiatry, 53*(1), 27–36.

Taylor, S. M., How, J., Travers, M. J., Newman, S. J., Mountford, S., Waltrick, D.,... & Gaughan, D. J. (2022). SMART drumlines ineffective in catching white sharks in the high energy capes region of Western Australia: Acoustic detections confirm that sharks are not always amenable to capture. *Biology, 11*(10), 1537.

Thiele, M., Mourier, J., Papastamatiou, Y., Ballesta, L., Chateauminois, E., & Huveneers, C. (2020). Response of blacktip reef sharks *Carcharhinus melanopterus* to shark bite mitigation products. *Scientific Reports, 10*(1), 3563.
Vagle, S., Chandler, P., & Farmer, D. M. (2005). On the dense bubble clouds and near bottom turbulence in the surf zone. *Journal of Geophysical Research: Oceans, 110*(C9).
Whitmarsh, S. K., Amin, D. B., Costi, J. J., Dennis, J. D., & Huveneers, C. (2019). Effectiveness of novel fabrics to resist punctures and lacerations from white shark (*Carcharodon carcharias*): Implications to reduce injuries from shark bites. *PLoS ONE, 14*(11), e0224432.
Woolgar, J. D., Cliff, G., Nair, R., Hafez, H., & Robbs, J. V. (2001). Shark attack: Review of 86 consecutive cases. *Journal of Trauma and Acute Care Surgery, 50*(5), 887–891.

Chapter 8
Paddling Against the Tide: Surf Activism in the Age of Plastic

Jennifer Savage

If the act of surfing is meant to bring joy, how bad does the environment around the surfer have to get before that joy dissipates? Let's say the waves are rolling in one after the other, fun-sized and glassy, with plenty of peeling rights and lefts for you and your pals in the water. A total stoke-fest. And then you see a plastic water bottle bobbing along. A bummer, sure, but probably not enough to ruin your day. Then you see a couple plastic bags, a candy bar wrapper. Maybe seeing the way the filmy plastic undulates on the surface takes you out of the moment by reminding you that plastic pollution is harming fish and birds who mistake it for food. But another wave comes and you move on. What if the amount of plastic in the water can't be escaped, however? What if you're pushing trash away from you while paddling, seeing it rising up in the wave face, watching as a seal bobbles by with pieces of plastic draping off its fur? Is surfing still fun?

My answer is no. A photo of a surfer in Bali catching a wave suffused with plastic trash made the rounds a while back—and islands do bear the brunt of a society used to throwing stuff "away"—but I've exited places in my home state of California too overcome with disgust to ever paddle out there again. I think of my last session at Ray Bay, a place where the San Gabriel River empties into the Pacific, and although I'm pretty sure I caught several sweet little longboard waves, what I mostly remember is how the garbage surrounded us, washed from the city streets down the concrete-walled waterway into the lineup and beyond. Most of us would refuse to surf waves polluted with toxic bacteria and obvious sewage, rightly afraid of getting sick. We should think that way about plastic, too.

More than an unsightly annoyance, plastic contains more than 16,000 types of chemicals, with at least 4200 of those considered to be "highly hazardous" to human health. Anyone who's dipped into the headlines in recent years knows that plastic breaks down and down and down without ever disappearing and those micro particles

J. Savage (✉)
Surfrider Foundation, Manila, CA, USA
e-mail: jsavage@surfrider.org

L. Mach (ed.), *The Future of Surfing in the Anthropocene*,
https://doi.org/10.1007/978-3-032-09377-6_8

have made their way into places we hold dear: the spray of waves, the breath of dolphins, our own lungs.

The fact that microplastics have penetrated our bodies—in addition to the lungs, they show up in our brains, hearts, blood, reproductive organs, livers, skin—and the most remote places on Earth, from Mount Everest to the Mariana Trench, does not inspire optimism toward a happy ending. So, before I go any further, know this: Plastic pollution is a solvable problem. We can't undo what harm has been done nor the harm unfolding as I type, but we can fix the future by forcing change now.

Why So Much Plastic?

I can't escape plastic pollution. If I'm not returning home from a litter-laden surf session or a particularly brutal beach cleanup, I'm immersed all day in work emails, new stories and policy efforts related to the impacts of plastic trash on our world. Even though I know that eliminating plastic pollution requires corporate and governmental change, seeing the harm firsthand and knowing about the related suffering happening in distant places makes me not want to be part of the problem. What to do?

Our reliance on single-use plastic and the resulting consequences stems from two main sources: the fossil fuel industry's never-satiated hunger for profit and the very human attraction to convenience.

Of note: People can, and do, make arguments for certain applications of plastic as necessary such as medical equipment, transportation, building materials. Plastic manufacturers tout healthcare products in particular as a reason to push back against any proposed regulation, arguing that, "Plastic saves lives," while failing to acknowledge that the production of plastic is so toxic to the communities around petrochemical factories, such as Louisiana's "Cancer Alley," that those areas are commonly known as "sacrifice zones" (Columbia University School of Professional Studies, 2023). In contrast, organizations like Healthcare Without Harm call for eliminating toxic chemicals from the products and places dedicated to saving lives. Discussions about how to more safely make "necessary" plastic are important, but should not be used to distract from the less complex truth that ending plastic pollution and all its harms requires that unnecessary single-use plastic production must end.

The Fossil Fuel Connection

And yet, manufacturers expect to produce *twice* as much plastic by 2040 and *three times* the current amount by 2060. This is largely because plastic is made from crude oil and natural gas and, as shifting policies cause petrochemical companies to experience a reduction in energy profits, those same corporations see increasing plastic production as the best way to keep the money flowing in. In addition to the production of plastic being an inherently toxic process, the effort to extract the fossil fuels necessary to create plastic products emits more greenhouse gases than almost all countries in the world. For context, if the plastic industry was a country, it would be the world's fifth largest greenhouse gas emitter (Environmental and Energy Study Institute, 2021). As plastic deteriorates, it releases methane and ethylene, drastically worsening climate change impacts and increasingly poisoning our water, air, food, and bodies with microplastics.

The connection between oil companies and the pushing of plastic products onto the global population has grown even more evident in recent years. In September 2024, a coalition of non-profit environmental organizations including the Surfrider Foundation, the Sierra Club, Heal the Bay, and San Francisco Baykeeper, represented by Cotchett, Pitre & McCarthy, filed a lawsuit against ExxonMobil (Exxon), the world's largest producer of single-use plastic polymers. The suit alleges Exxon deliberately concealed the harms caused by single-use plastics and systematically led the public to believe that plastic waste is easily and safely disposable via recycling, incineration or landfills.

In reality, only about five percent of plastic is recycled and the rest of it persists in the environment for thousands of years leaking toxic chemicals that are known to harm human health, wildlife and habitat. As a result, the lawsuit alleges, Exxon's conduct has directly worsened the plastic pollution causing harm to California's coasts and waterways. California Attorney General Rob Bonta filed a similar action the same day. In return, in January 2025, Exxon filed a lawsuit about Attorney General Bonta and the coalition of environmental groups, claiming the legal battle was part of a "smear campaign."

What's a Person to Do?

On the consumer front, plastic often lines every inch of a store's shelves, from beverages to detergents to tortillas to vegetables bagged in plastic. Toothbrushes, shoes, the tags hanging from clothes, the clothes themselves. Prepared food, every toy in the toy aisle, office supplies, makeup, skincare products. The saline solution I need for my contacts. Party supplies. Take-out food. The ubiquity of plastic has the effect of normalizing the weirdness of buying something that will be used once but last forever. Even when we want to avoid it, alternatives don't always exist or, when they do, the upfront costs are discouraging. We've grown so reliant on single-use plastic, imagining a world without it—which is to say, the world as it existed up until about 80 years ago—seems impossible.

Current Conditions in the Lineup and Beyond

Imagine a line of garbage trucks rolling up to the beach, trundling out on the sand and unloading a full load of plastic trash into the ocean every minute of every day of the year. That is an accurate vision of how much plastic—11 million metric tons—enters our ocean worldwide (United Nations Environment Programme, 2021). If nothing changes, that amount will double every six years. If the industry gets its way, that amount will be exponentially higher.

People don't want this. More than 90% of U.S. consumers consider the amount of plastic in a product before deciding on a purchase. An average consumer might chuck their yogurt containers and soda bottles into a blue bin once the food has been eaten and the beverages drunk, but even with recycling programs in place, in the U.S. less than five percent of our plastic waste is recycled into another plastic product. Even when plastic is recycled, it is often just recycled once and then discarded into landfills, incinerated or dumped directly into the environment. This is why Exxon and other plastic manufacturers continue to scale up their virgin plastic polymer production capacity without regard for the harm they're perpetuating on all of us.

Researchers estimate that between 50 and 75 trillion pieces of plastic and microplastics are in the ocean already (UNESCO, n.d.). Some of the problems caused by this pollution are obvious ones. Think of the stories about starved whales found with dozens of plastic bags in their bellies, the seals strangled by plastic rings, the turtle bloodied by the straw embedded in its nostril. Those are only the stories we see. Cigarette butts, the *most littered item in the world,* are so toxic that a single cigarette butt in a liter of ocean water is lethal to marine life. Expanded polystyrene foam products, including foodware, break into small pieces that litter the beach and poison marine species when ingested (in addition to releasing toxic chemicals into your hot takeout food during use). Microbeads and microfibers are found in the stomachs of fish, then bioaccumulate up the food chain. Something to consider as you're eyeing the sushi menu—how many toxic chemicals do you feel like eating in your ahi roll? Scientists have found that dolphins are emitting plastic particles as they exhale and waves are spraying microplastics into the atmosphere (Dziobak et al., 2024). How does one feel joy in the lineup knowing this means we are breathing in plastic, too? What happens to the soul-cleansing connection with nature when that connection is coated in plastic?

Life doesn't have to be like this. We do not need to sacrifice our planet, our ocean, our very bodies as if no other options exist. We can demand of our lawmakers that they act on behalf of those they represent and start regulating plastics like the toxic products they are. We can strive to hold true to our own values in our own lives, whether that's bringing our own bag to the store or taking the best care of what we own to ensure it lasts—particularly important for surfers given most of the equipment we use, from boards to wetsuits, is made of plastic.

Change is Possible

On that last note, efforts to make more sustainable equipment continue to expand. We've seen innovations in making boards with recycled materials including using 3-D printing and shapers utilizing various types of wood to better align their respect for the ocean with their love of riding waves. Patagonia's commitment to Yulex wetsuits has pushed larger manufacturers to commit to a shift away from neoprene (Zheng et al., 2023). Regardless of what type of board you ride, ensuring it will last as long as possible is key—"repair instead of replace whenever possible" is a useful motto for everything you own.

Individual actions like these serve to normalize sustainability in a disposable-oriented culture. They allow us to improve our lives by reducing the plastics within our own homes. Hopefully, incorporating more thoughtful ways of existing also inspires us to support organizations like the Surfrider Foundation fighting to make the policy changes critical to stopping plastic pollution at the source.

How Good Laws Stop Bad Actions

Of course, that will require fossil fuel corporations and other companies that profit off plastic to cut back on production—and that will likely require new laws, policies, and international agreements. Over the past several years, multiple members of Congress have introduced legislation aimed at solving the worst aspects of plastic pollution. The most significant of these bills is the Break Free From Plastic Pollution Act, first introduced in 2021, reintroduced in 2023, and expected to be brought forth again in 2025. BFFPPA, as it is known to supporters, is the most comprehensive federal legislation ever written to address the toxic plastics crisis and protect people and the planet from corporate polluters. Unfortunately, the Trump Administration's focus on promoting fossil fuels and determination to retreat from climate change solutions, along with the acrimonious divide between political parties in Congress, means anti-pollution activists are desperately trying to hold the line with little hope of advancing the fight against plastic waste.

One legislative bright spot in the past few years—the Plastic Pellet Free Waters Act—actually scored Republican support after introduction, largely due to the focus being extremely narrow and relevant to those living in a state where a train derailment spilled millions of plastic pellets into the environment. These types of plastic pellets, also known as "nurdles," are the pre-production building blocks of nearly all plastic goods. Unfortunately, while a train derailment might make the news, the fact is that current law allows plastic manufacturers and shippers to discharge trillions of these plastic pellets directly into America's waterways and ocean with little to no enforcement despite the huge risks to our ecosystems these tiny granules of plastic pose. According to a 2020 report, an estimated 200,000 metric tons of plastic pellets (approximately 10 trillion nurdles!) make their way into the ocean every year. They are often mistaken for fish eggs or other food by sea life and birds, and can lead to malnourishment and death.

The Plastic Pellet Free Waters Act would have required the U.S. Environmental Protection Agency to prohibit the discharge of plastic pellets or other pre-production plastic materials from facilities and sources that make, use, package, or transport those materials. While the bill, like most proposed in an increasingly ineffective Congress, didn't pass, efforts to address the pellet problem continue at the state level.

In 2024, Surfrider met with more than 140 Congressional offices to educate elected officials and their staff about the harm being done by plastic pollution, including by microplastics and nurdles. These efforts will continue, as will our work to pass effective local and statewide ordinances. As noted above, continuing to build momentum at the grassroots level is likely to prove most effective in making change.

In the Streets and in the Lineup, Grassroots Activism is Where to Find Hope

In Surfrider's 40 years of advocating for our ocean, waves, and beaches, we've seen time and time again how local successes scale up to larger ones. The forecast is promising, with momentum building: more than 500 local and state bag laws have been adopted in the U.S. Data from Surfrider's Healthy Beaches beach cleanup program helps make the case for why such laws are necessary. "Plastic Bag Bans Work," a 2024 report from Environment America, U.S. Public Interest Research Group Education Fund and Frontier Group showed that statewide bans on plastic bags in New Jersey, Philadelphia and Vermont, along with bans in Portland, Ore. and Santa Barbara, Calif., reduced the number of single-use plastic bags used each year by around 6 billion. If you go to the beach somewhere a ban on single-use plastics has been implemented, you'll often notice you see less of that kind of trash—plastic bags, straws, etc.—being washed or blown from the streets to the surf. (This doesn't address the problem of trash washing up from elsewhere, unfortunately!).

Meanwhile, Surfrider's Ocean Friendly Businesses programs provide restaurant and hotel owners choosing to operate sans single-use plastics an array of resources and promotional support. Likewise, business owners are a particular effective constituency when making an economic argument to advocate for what can be done and pressure lawmakers to follow suit.

As more cities pass laws banning single-use plastics and implementing reuse practices, achieving the same goals at the state level becomes reasonable. As more states embrace legislation to reduce plastic pollution, we move closer to national success within the U.S. Given our country's outsized impact—we produce more plastic waste than any country in the world—making change at home will offer the entire world hope for a plastic pollution-free future. And even if the U.S. fails to step up, the rest of the world won't wait. The fossil fuel industry's lobbyists vastly outnumbered environmental and social justice advocates as members of the United Nations attempted to negotiate a global plastic treaty over the past few years, effectively delaying it, but more than 100 countries advocated for real solutions, including the reduction of plastic production (Center for International Environmental Law, 2024). Believing progress is being made can be difficult when so much plastic pollution still exists in the ocean, on our beaches and along our streets; nonetheless, every time we choose to skip the single-use plastic options, every time we show up to voice support for a healthier, plastic pollution-free planet, we're one step closer to restoring our ocean to a place of joy and hope.

References

Center for International Environmental Law. (2024, March 5). *Lobbyists at INC-5: Who is shaping the global plastics treaty*? https://www.ciel.org/news/inc-5-lobbyist-analysis/

Columbia University School of Professional Studies. (2023, October 11). The endless shame of Louisiana's "Cancer Alley". *Columbia SPS News*. https://sps.columbia.edu/news/endless-shame-louisianas-cancer-alley

Dziobak, M. K., Fahlman, A., Wells, R. S., Takeshita, R., Smith, C., Gray, A., Weinstein, J., & Hart, L. B. (2024). First evidence of microplastic inhalation among free-ranging small cetaceans. *PLoS ONE, 19*(10), e0309377. https://doi.org/10.1371/journal.pone.0309377

Environmental and Energy Study Institute. (2021, December 9). *Plastic production and climate change*. https://www.eesi.org/briefings/view/120921waste

United Nations Environment Programme. (2021). *From pollution to solution: A global assessment of marine litter and plastic pollution*. https://www.unep.org/interactives/pollution-to-solution/

United Nations Educational, Scientific and Cultural Organization (UNESCO). (n.d.). *Plastic pollution in the ocean*. Ocean Literacy Portal. https://oceanliteracy.unesco.org/plastic-pollution-ocean/

Zheng, J., et al. (2023). Plastic waste and climate change: A growing global challenge. *PNAS Nexus, 2*(10), pgad296. https://academic.oup.com/pnasnexus/article/2/10/pgad296/7286531

Chapter 9
To Surf or Not to Surf: Risk Tolerance in Polluted Blue Spaces

David L. Brown

See (Fig. 1).

Total Submersion

Surfing involves total submersion into the water. You can't go surfing without getting water in your eyes, ears, or mouth, as hard as some of us may try. Every one of our five senses is activated for the entire time we're out in the water. Feeling the offshore wind running through your hair, hearing bubbles rush past your ear telling you which way is up after a wipeout, smelling and tasting the saltwater during that first duck dive while paddling out, or watching the sun rise or set over the horizon to kick off or end a session. Through this total immersion, surfers become addicted, and this passion turns into an insatiable itch that can never be fully scratched.

Like many of us, I learned to surf with my family. My dad first put me on the nose of his longboard when I was only two years old. I grew up in landlocked Arizona, so I only had the opportunity to surf when we drove out to California. I quickly learned on those trips to adopt an "ignorance is bliss" mindset; if there were waves to be ridden, I jumped in no matter what, regardless of poor conditions or polluted water. When I moved to California full time for college, I was perpetually sick from winter surfing. After my third sinus infection in as many months and being on a first name basis with the health clinic staff, I finally started to heed rain advisories.

Given my affinity for the water, I chose to study Environmental and Ocean Sciences during undergrad to better understand my home away from home. However, once graduating in 2020, employment opportunities in the midst of the COVID pandemic were slim. One of the few sectors exempt from hiring freezes was public

D. L. Brown (✉)
Independent Researcher, San Diego, CA, USA
e-mail: davidbrown@sandiego.edu

L. Mach (ed.), *The Future of Surfing in the Anthropocene*,
https://doi.org/10.1007/978-3-032-09377-6_9

Fig.1 A surfer enters the ocean after a 25,000-gallon sewage spill at North Pacific Beach, San Diego on March 9, 2024

safety. I threw my name in the hat, and after completing the San Diego Regional Lifeguard Academy, I began working the beaches as a San Diego Fire-Rescue Department seasonal lifeguard. I fell in love with being paid to work at the beach every day and making a difference in the community by saving lives. As an industrial stormwater scientist and inspector during the other eight months of the year, I was one of the few who possessed the analytical knowledge to understand the water quality we were making rescues in. After several years, I began growing concerns about the ocean water conditions I was immersing myself in almost every day to surf, bodysurf, swim, dive, boogie, connect, relax, and reset. I shifted paths to become a full time Stormwater Environmental Specialist with San Diego's Stormwater Department, a position I still hold.

In the past few years, I've already experienced two "once in a lifetime" weather events and seen friends impacted by a third. I witnessed Tropical Storm Hillary impact the San Diego coast on August 20th, 2023, firsthand while sitting in a fiberglass lifeguard tower on the beach violently shaking under the strain of 50+ mph gusts of wind. Only five months later a 1000-year storm struck San Diego prompting over 250 water rescues throughout the city from rapidly rising and heavily polluted flood waters conducted by lifeguards, firefighters, and community members. Devasting downpours turned into an unprecedented drought. This was coupled with the hottest summer and one of the dryest wet seasons in recorded history, which culminated in the firestorms that decimated Los Angeles. These compounding disasters, like a river draining to the sea, meander through all walks of life, identities, and socioeconomic backgrounds. They trickle down until reaching a cascading flow raging into the sea.

Sitting in a little fiberglass tower by myself staring out at the ocean for hours on end gave me plenty of time to think, for better or for worse. Lived experience is a powerful thought provoker and opened my eyes to the increasing interconnectedness of everything in life. As we move further into the Anthropocene, in many ways, the future of surfing will look vastly different than it did for generations of surfers prior. However, the core motives of leisure, relaxation, spirituality, comradery, and chasing the endless tube will endure. To ensure the art of wave riding continues long after we're gone, we must take a more active role in preserving and protecting the sacredness of the ocean for the generations of surfers still to come. In this chapter my goal is to share my experiences and what I have learned from experiencing 'unprecedented' events, studying storm water, lifeguarding, and surfing in Southern California.

Polluted Leisure and Surfer Health

The World Health Organization (WHO) defines health in their preamble as 'a state of complete physical, mental and social wellbeing and not merely the absence of disease or infirmity' (World Health Organization, 1946). Many of us use surfing as an outlet to not only improve physical health, but mental health as well (Buckley & Cooper, 2023; Gibbs et al., 2022; Mach et al., 2024; Olive et al., 2023). Floating in a seemingly endless body of water allows us to decompress from the stressors of life on land. Furthermore, there's nothing like being thrown over the falls and throttled to the point of exhaustion to make one realize the important things in life. Physical immersion in blue spaces such as the ocean provides an immense calming relief for our mind, body, and soul (White et al., 2010) as well as numerous mental health benefits (Olive & Wheaton, 2021). Consistent evidence demonstrates positive correlations between regularly engaging with blue spaces and improved overall health and wellbeing (Britton et al., 2020; White et al., 2020).

While I like to think of surfing as a mental health necessity, others may call it an unhealthy obsession. Surfing will often take precedence over all other activities whether it be work, social gatherings, or writing this chapter. So many of us utilize surfing as a mental health coping mechanism whether we are conscious of it or not, but surfing shouldn't jeopardize our physical health in the process.

The quality of the water that we surf in and enjoy affects our physical health. Clifton Evers (2021, pg. 2), in his work on polluted leisure, discussed how "pollution can complicate any positive correlations between nature-based sports, blue spaces, and health/well-being." To Evers (2021, pg. 2) polluted leisure means "the embodied, sensorial, emotional, intellectual, spatial, and technological busyness of pollution." Pollution permeates all aspects of life, including spaces of leisure meant to be used as an escape, source of enjoyment, and decompression. As coastal regions around the world have become increasingly urbanized, the rate of pollution entering waterways and the ocean has increased as well (Freeman et al., 2019; Strokal et al., 2021; Xu & Zhang, 2022). In a sense, this means we cannot actually escape pollution in many of our leisure pursuits, we can only really find ways to cope with its ubiquity.

There's no better example of this than where I live and work in California, where the population has exploded from 2.3 million people in 1910 to nearly 40 million in 2020, a 1639% increase in the last 110 years (U.S. Census Bureau, 2021). The majority of California's nearly 40 million people live within the 15 coastal counties bordering the Pacific Ocean. All these people inevitably generate a lot of waste that finds its way into the ocean. Impervious surfaces within the built environment like roads, sidewalks, and roofs generate more pollutants and runoff than natural landscapes that allow for infiltration and filtration of runoff (McFarland et al., 2019). Watersheds with greater impervious cover exhibit higher pH, specific conductance, and temperature, along with elevated nutrient loads, total phosphorus, and bacterial concentrations (fecal coliform and *E. coli*) and increased maximum runoff flow (Chelsea et al., 2012; Schueler, 1994; Yabsley et al., 2020). Epidemiology studies have shown that exposure to recreational waters contaminated with Fecal Indicator Bacteria (FIB) from wastewater and urban runoff correlates with an increased risk of gastrointestinal illness, respiratory disease, ear infections, and skin infections (Arnold et al., 2016; Colford et al., 2007; Haile et al., 1999; Tseng & Jiang, 2012; Wade et al., 2006; Yau et al., 2009, 2014).

Poor coastal water quality has amounted to 3.56 million sick leave days per year across California due to recreational beach water pollution (Li & Zhang, 2019) as well as an economic loss of at least $51 million in Los Angeles and Orange Counties alone (Given et al., 2006). Meanwhile, DeFlorio-Barker et al. (2018) estimated 90 million illnesses, with an economic cost of $2.2- $3.7 billion annually nationwide. More regions and surf breaks around the world will continue to be impacted by decreased water quality due to urban coastal development. These issues will be most apparent in underdeveloped regions of the world as surf tourism continues to expand into previously "remote" surf destinations. Without adequate forethought and planning of critical infrastructure that's suitable to adapt to increasing populations and climate change, water quality impacts will be felt heavily (Buckley, 2002; Ponting, 2009).

For most of this chapter, I will focus on coastal water quality within the United States and specifically in California. However, much of the information presented about bacterial, algal, urban, and agricultural water pollution can be extrapolated across the world. Essentially any pollutant generating activity on land eventually finds its way downstream into the ocean. Governments and regulatory bodies love to create acronyms for just about everything, leading to alphabet soup that can be difficult to follow. Throughout this chapter all acronyms will initially be written out in full while subsequent mentions will utilize their respective acronyms.

The Politics and Techniques for Understanding and Monitoring Water Quality in the Surf

The following paragraphs are going to provide a broad overview of water quality monitoring within the United States and the analytical testing methods utilized. Some of the content is complicated, but I promise the whole chapter won't just be a bunch of technical jargon. I've done my best to pare the following analytical information down to its bones. Unless you're a molecular biologist, I don't think many of us fully understand how these different tests work. I felt it was important to include as much detail as possible for those who are curious about the specifics of water quality monitoring at the beaches where we surf.

In an effort to combat high levels of illness causing bacteria in recreational waters around the United States, the Beaches Environmental Assessment and Coastal Health Act (BEACH Act) was signed into law in 2000. The BEACH Act amended the Clean Water Act (CWA) to ensure monitoring and public notification of contamination for all coastal recreational waters throughout the United States. Federal funds are dedicated to support recreational water quality monitoring across 39 states, territories, and tribes (U.S. EPA, Beach Management Resources, 2024). The task of sampling, testing, and notifying the public largely falls on a patchwork of county and tribal health departments instead of a uniform national system. While Congress authorized $30 million a year to be dedicated to the BEACH Act, only approximately $10 million a year has been granted to the program since 2000, leading to significant public health shortcomings (Dias, 2024). The already meager funding allocated to the BEACH Act has routinely been on the chopping block to "streamline government expenses" or to "trim the fat". All federal funding was set to be eliminated for the grant program in 2013, 2014, 2015, and 2019 (Beachapedia, 2018). However last-minute bipartisan congressional support and community advocation from groups like the Surfrider Foundation ensured the program wasn't cut. Due to a lack of sufficient funding, many local health departments only monitor beaches and recreational waters seasonally, such as during summertime when more people are in the water. This leaves a large knowledge gap for water users, especially surfers, who are in the water year-round, as opposed to specific seasons.

Health departments collect water samples via grab method. This is done by attaching a bottle to the end of a long pole and sweeping it just below the water's surface in at least knee-deep surf. Grab sampling can be labor-intensive, time-consuming, and costly because staff need to collect samples from multiple beaches located in different areas. The samples are taken to a public health lab where they are tested for fecal indicator bacteria; either *E. coli* or Enterococcus, or both. These specific bacteria are used to indicate the presence and level of fecal contamination in water (Krawczyk et al., 2021). High levels of *E. coli* or Enterococcus indicate an increased risk of sickness through mucus membrane exposure (Fleisher et al., 2010). Sample analysis is conducted utilizing one of three approved Environmental Protection Agency (EPA) methods: culture-based, quantitative Polymerase Chain Reaction (qPCR), or digital droplet Polymerase Chain Reaction (ddPCR) (U.S. EPA, OFFICE

OF WATER 820-F-058, 2012). While all three methods are approved, most departments employ the culture-based method where live bacteria colonies are grown on a petri dish and then counted by hand under a microscope. Due to its reliability, simplicity, and cost effectiveness the culture-based method is usually the preferred standard. Culture-based results generally take 18–96 h to incubate before they're able to be read while qPCR results come back in as little as two hours (Noble et al., 2010). The qPCR method is in use in Florida and the Great Lakes but costs nearly five times as much as the culture-based method and requires more complex steps to complete. Since there are a couple hundred coastal counties in the United States, federal grant dollars through the BEACH Act dry up quickly. As a result, funding for these monitoring programs must compete with other county programs that have a higher priority, such as public safety or infrastructure maintenance. When it comes to making monitoring program decisions, governments usually choose the cheaper option, unless legally mandated otherwise, even if it comes at the expense of the public being better informed and protected.

The most comprehensive, but also the most expensive analytical method to date, known as digital droplet Polymerase Chain Reaction (ddPCR) was recently approved by the EPA in 2022 to be used for beach monitoring in San Diego County. San Diego County became the first Federal and State approved coastal county in the nation to begin using the new ddPCR method. The method provides more rapid results to inform the public of the water quality conditions within the same day. Its usage will serve as a pilot program for the rest of California and likely, eventually the entire BEACH Act program. The new ddPCR method amplifies bacteria DNA in water samples to measure both live and dead FIB levels whereas the culture-based method only measures live bacteria. The ddPCR method is the most precise test to date due to its heightened sensitivity but is still in the early days of being utilized for beach monitoring and will likely need continued refinement before being implemented state of nationwide.

Beach specific water contact advisories are attributed to individual grab samples collected by health departments, but broad advisories are also utilized. General Rain Advisories are issued when rainfall equal to or greater than 0.20 inches is received along the coastal areas (DEHQ, 2024). This quantity of rain is the standard due to its ability to produce runoff from impervious surfaces which leads to the transport of bacteria and pollutants into stormwater drainage systems which eventually lead to the ocean. Health departments advise beach users to avoid contact with ocean and bay water for a period of 72 h after rainfall ends to allow for FIB levels to return to pre-rainstorm levels, which generally diminishes within three to five days (Ackerman & Weisberg, 2003; Leecaster & Weisberg, 2001). Some bacteria such as Enterococcus have the capacity to live outside of a host and cause infection for approximately three days (Rozen & Belkin, 2001; Fujioka et al., 1981). However, more robust pathogens such as Norovirus can live up to three weeks in ocean water without a host (Rexin et al., 2024). Luckily due to a variety of factors such as tidal fluctuations, sunlight, winds, currents, and surf most pathogens are diluted enough to not pose a threat to human health after three days.

Every year, the EPA releases a report summarizing the monitoring results from the BEACH program and provides data analysis of potential sources of contamination. The annual report on coastal recreational water quality paints an overall positive picture of the 3561 monitored beaches nationwide. Monitored beaches throughout the United States were open and safe for swimming or surfing 91% of the time in 2023 (U.S. EPA, Beach Report: 2023 Swimming Season, 2024). Unfortunately, only 70% of the beaches under the BEACH program are monitored for bacteria. Therefore, smaller beaches and stretches of coastline that are difficult to access (but provide tasty waves) are overlooked, and therefore may be subject to unhealthy bacteria levels. Resources are stretched thin due to lack of funding, so efforts are focused where the highest number of people will come into contact with the water, a phenomenon most surfers are averse to. Some states have a more widespread monitoring system than others, and they are often clustered around large population centers. Massachusetts has largest number of monitored beaches with 574, followed by New York (338), Florida (263), and California (252) (U.S. EPA, Beach Report: 2023 Swimming Season, 2024). Monitoring programs are required to list a possible pollution source when water quality advisories are issued. More than a third (41%) of the pollution sources were reported as unknown, likely due to source identification not being a covered cost under the previous iteration of the BEACH Act, while the second most reported source (24%) was runoff (U.S. EPA, Beach Report: 2023 Swimming Season, 2024). Depending on where you live, rainfall may be continual throughout the year. In the case of Southern California, rainfall is concentrated during one season, winter. In these dry coastal places, throughout most of the year, stormwater infrastructure including streets, curbs, levees, pumpstations and storm drains remain dry. While they may be dry, that doesn't stop pollutants like oil, grease, trash, dog poop, decaying organic matter, nutrients, pesticides, heavy metals, and human fecal matter from accumulating in the system. All that and more builds up, just waiting for a rollercoaster ride to the ocean with the first major storm of the season, unaffectionately known as the *first flush*. The first flush of the stormwater system for the wet season is also the most toxic, to human, wildlife, and overall environmental health (Bertrand-Krajewski et al., 1998; Katz et al., 2006). This massive accumulation is a significant source of bacteria and other illness causing pollutants that are discharged into coastal waters all at once. Stormwater runoff discharges carry a whole host of pathogens, agricultural, and industrial waste into the lineup that can cause adverse health impacts on those of us who turn to the ocean for fun and livelihood.

Risk Versus Reward and the Desire to Score

The art of wave riding and surfing dates back thousands of years as an expression of joy, love, connectedness, and spirituality. Through a soulful bond with the ocean, surfers have proven time and time again to be some of the most ardent, resilient, and dedicated individuals to the sea. Surfers are often the first outsiders to return to regions after armed conflict and social unrest (Laderman, 2014; Old,

2020); terrorist attacks (Gurtner, 2004; Robinson & Jarvie, 2008); epidemics (Wavelength, 2016); and natural disasters like the tsunami and resulting nuclear meltdown in Fukushima in 2011 (Evers, 2021). They're driven by the insatiable urge to ride the best waves while vying for priority with the fewest number of people. These favorable conditions are next to impossible to find in Southern California, except for during periods of inclement weather. This inclement weather generally coincides with large, northwest swell producing storms from the North Pacific. These storms often bring high pollutant generating rainfall rates along with large waves that surfers prefer (Bradley & Hancock, 2003; Steele et al., 2018). Such events force surfers in urbanized environments to undergo a personal risk versus reward evaluation.

The risk versus reward model of choosing whether to risk illness through exposure to polluted water or score waves looks different for each individual surfer. Most leisure is polluted to some degree and individuals must determine personal measures to mitigate and reduce exposure as much as possible (Evers, 2021). Some base their choice on whether to surf in polluted waters on geographical factors, such as proximity to pollution sources, like storm drains, while others adopt an ignorance is bliss mindset and prefer to operate under the "what I don't know can't hurt me" model. Surfers in the Gulf of Maine indicated that 47.1% of them factor water pollution into their decision-making process prior to paddling out for a session (Scott, 2018). Other geographical factors like land use along the coast plays a significant role in the volume of pollution impacting surf breaks (Dwight et al., 2002). A study comparing urban built environments in North Orange County to rural environments in Santa Cruz County identified a correlation in resulting illnesses due to runoff exposure (Dwight et al., 2004). While Santa Cruz received more rainfall than Orange County during the study period, surfers in Orange County reported higher rates of every symptom. Illnesses reported by surfers following heavy rainfall included eye redness, skin infection, sinus problems, nausea, stomach pain, cough, vomiting and diarrhea (Dwight et al., 2004; Harding et al., 2015). Even if there hasn't been a recent rain event, or an identified point source, like a sewer main break, risks of illness are still involved. Findings indicate that ocean goers are inherently at a higher risk of getting sick due to exposure to sea water when compared to non-ocean goers (Fleisher et al., 2010; Leonard, 2018; O'Halloran & Silver, 2021). The risk of illness from surfing is an ever-present underlying risk but increases when polluted runoff enters the picture.

Many health studies focused on surfers have not collected real time water quality data to correlate with self-reported illnesses. However, a first of its kind study, The Surfer Health Study, conducted by Southern California Coastal Water Research Project (SCCWRP) scientists in 2015, funded by the City and County of San Diego, has helped fill the gap tying water quality data to illness. The study lasted through two wet seasons and surveyed 654 surfers from two locations in San Diego: Tourmaline Surf Park and Ocean Beach. Water samples were collected upstream in Tourmaline Creek and the San Diego River before runoff entered the ocean, as well as ocean samples at the respective surf breaks. Surfers self-reported every surf session, the duration, and any subsequent illnesses on a weekly basis. Illness rates were calculated to mirror the EPA's recreational water criteria for estimated illness rates per 1000 swimmers, or surfers in this case. Illness rates increased during the 72 h window after rainfall when a public health department general rain advisory was in effect. Any infectious symptom was the most reported illness category with 47 separate cases during wet weather which correlated to a rate of 16.1 illnesses per 1000 surfers. Ear infection (12.3/1000), sinus infection (12.3/1000), upper respiratory infections (12.2/1000), and gastrointestinal illness (10.2/1000) saw the largest increases during wet weather when compared to dry weather (Arnold et al., 2017). Additionally, illness rates peaked within 0–24 h after rainfall occurred, demonstrating that the ocean was the most contaminated within the initial 24 window after rainfall. Finally, bacteria concentrations at beaches were highest directly in front of stormwater discharge points, but tapered off the farther a sample was collected from a discharge point (Schiff et al., 2016). In an artform that requires numerous variables to align to create a wave just long enough to be ridden, it makes sense that even more factors determine water quality.

Going with the Flow: Long Range Transport of Inland Pollutant Generating Activities

I vividly remember surfing with my dad on his white longboard as a kid during a red tide one summer; I was convinced that the red tide was blood in the water from a shark feeding frenzy. At six years old, no one could tell me any differently. I was so scared that I made my dad take me back to shore where I firmly stayed for the rest of the day. I was terrified for everyone else still out in the water that the men in the grey suits were going to pick them off one by one when they kicked out from the peeling rights. To my relief, my dad and everyone else made it back in safely after hours of scoring impeccable waves. My first experience with a red tide started as one of fear but has since evolved to one of curiosity about what causes them and how they spread. Concern about red tides seems to be a common line of thought among many people. As a lifeguard, we would receive dozens of questions every day during a red tide, "Is it safe to swim in; Does it attract sharks (see, it's a reasonable suspicion); Will my kids get a sinus infection (maybe); We just got into town, will this last our entire vacation; Why does it smell so fishy; Can I put my head underwater; Will it stain my boardshorts; Can you make it go away; Are there other beaches nearby without it?" While one of the most important tasks as a lifeguard is to be an ambassador for the ocean, during those days it felt like we were a magic eight ball. Some days the water would be crystal clear in the morning but once the onshore winds started in the afternoon, the red tide would creep back in, and visibility would drop to zero. We didn't always have the answers, although some may seem obvious. The ocean is one big mystery which draws so many of us to it and to study it.

Luckily there are researchers out there investigating the phenomenon of red tides and answering the big questions for us, like will it stain my boardshorts? Dr. O'Halloran from UC Berkeley conducted a comprehensive meta-analysis and subsequent health study in 2011 on the impacts of red tides on surfers and lifeguards. Many of us in the surf community have heard the term red tide or harmful algal bloom (HAB) and use them interchangeably. However, they are different, red tides are a result of excessive growth of phytoplankton with reddish pigments and do not cause harm while HABs cause negative environmental and human health impacts by producing harmful toxins (O'Halloran, 2011). HABs bloom in coastal waters due to excessive nutrients like nitrogen and phosphorous from fertilizers washing off into the ocean and increased ocean temperatures due to climate change (Gobler, 2020; Moore et al., 2008). These HABs are steadily increasing globally and can last for weeks or even months and impact thousands of miles of coastline (Brand et al., 2010). On the west coast of the United States, a planktonic diatom called *Pseudo-nitzschia* produces a potent neurotoxin called Domoic Acid (DA) that can pose a major health risk for wildlife and humans alike (Gajski et al., 2024). Fortunately, the risk to surfers through dermal exposure or ingestion of small quantities of DA contaminated waters is minimal. The risk of illness from eating contaminated shellfish or getting bit by a sea lion gone mad due to the neurodegenerative effects of Domoic Acid Poisoning is another story all together. Researchers suspect that an influx of nutrient rich ash from the Los Angeles firestorms in January 2025 could potentially have contributed to the large HAB that impacted the entire Southern California coast in the spring of 2025. Meanwhile on the gulf and east coast, large volumes of fertilizers enter waterways from industrial agriculture facilities both along the coast and hundreds of miles inland. The Gulf of Mexico routinely sees HABs explode as thousands of tons of agricultural waste from Midwest states flows downstream through the Mississippi River. Due to an increase of coastal populations and anthropogenic pollution sources the toxic phytoplankton *Karenia brevis* saw a 20-fold increase in abundance within five kilometers of the shoreline from 1954 to 2002 in the Gulf of Mexico (Brand & Compton, 2007). Certain species like *K. brevis* can cause symptoms like coughing, sneezing, sore throat, shortness of breath, upper respiratory infections, rashes, and eye irritation after swimming in contaminated water (O'Halloran, 2011). Evidence from Dr. O'Halloran's analysis suggests that there is a positive association between HAB exposure and acute health symptoms in humans. These acute health symptoms primarily present in ocean users as upper respiratory infections, eye irritation, nasal congestion, skin infections, and cough.

Tijuana River Case Study: When Just About Everything Goes Wrong

The Tijuana Sloughs are a prime doomsday example of what could be without proper investment in the future. One of the premier big wave spots in California has been lost to environmental catastrophe. To be clear, the wave is still there in all its glory, unchanged since it was first ridden in the late 1930s (Wright, 1985). However, instead of hosting big wave competitions and being the backdrop for blockbuster surf movies, the lineup is a desolate ghost town. If the same wave was located literally anywhere else in California it would be crawling with hundreds of surfers itching to cut their teeth on the steep drop into a racetrack face.

Unlike most beaches where you're greeted with a pleasant sea breeze prior to paddling out, at the Sloughs, you're slapped in the face with the most heinous, eye watering, and gag reflex inducing smells. Gazing out at the lineup, the noxious sewage fumes may cause you to think you're hallucinating as perfect wave after wave rolls into shore. After a few minutes of disbelief, those toxic chemical induced tears give way to tears of joy watching the reeling faces rip over the cobblestone reef. However, the immaculate break is devoid of life, human and animal alike due to a plume of toxic brown soup spewing out of the Tijuana River year-round (Danil et al., 2023; Stigler et al., 2024). On average approximately 40–80 million gallons of raw untreated sewage enters the Pacific Ocean every day from Tijuana, Mexico (TJ) (Rico et al., 2025). Sewage from TJ enters the ocean through two main sources, the Tijuana River, and the San Antonio de los Buenos Sewage Treatment Plant, six miles south of the US border (Feddersen et al., 2021). If you've ever driven south down Mexican highway 1D to kick off a Baja surf expedition, you've likely hit a point early in the trip where someone in the car says, "maybe those street tacos weren't the best idea." Depending on the taco stand, they may be right, but hopefully the smell was just from the sewage plant discharging over 28 million gallons a day of untreated sewage via a waterfall underneath the highway (U.S. EPA, USMCA Tijuana River Watershed, 2025). The resulting water quality and health impacts from the sewage fluctuate based on swell and current direction. In the summertime, large south swells blow the sewage contaminated water north all the way up to Sunset Cliffs (SIO, UCSD, 2025), while during northwest winter swells the water becomes contaminated as far south as K-38s. At any given point throughout the year, nearly 30 miles of coastline is unsurfable due to potentially life-threatening pollution (Taylor, 2016).

According to the International Boundary and Water Commission (IBWC), since 2019, over 100 billion gallons of untreated raw sewage has spewed out of the TJ River into the Pacific Ocean (California Coastal Commission, 2023). After the sewage enters the ocean, the water becomes a source for waterborne disease resulting in chronic beach closures due to dangerous bacteria levels both north and south of the border (Stigler et al., 2024). Unfortunately, the responsibility of pollution mitigation is complicated and politically fraught. One might first jump to the assumption to solely blame Mexico where approximately 75% of the Tijuana River Watershed resides and a significant portion of the population does not have access to the sewer system. However, the situation, like water making its way downstream to the coast, is much more fluid. Industrial waste discharged into the Tijuana River from U.S. owned and operated manufacturing plants located in Tijuana known as *maquiladoras* contribute a significant volume of toxic pollutants (Johnsen, 2018; Honkanen, 2020). U.S. corporations set up *maquiladoras* where C suite executives and upper management are based in the United States, but manufacturing and working-class jobs are conducted just across the border (City of San Diego, 2025). These set ups are meant to maximize profits by skirting U.S. labor and environmental regulations implemented to protect worker, public, and environmental health. The toxic pollutants however, in mother nature's vision of karma, inevitably come home to roost and end up decimating coastal communities in the U.S. In a capitalistic economic model that values profits over everything, human and environmental health typically loses.

The Tijuana River crisis is a tri-national environmental justice issue impacting citizens of the Kumeyaay Nation, United States, and Mexico. Pollution is inherently transient; it isn't bound by political borders. It can be transported across many mediums and result in exponentially larger exposure profiles as it moves through the environment. Researchers have uncovered alarming quantities of both chemical and biological pollutants that can cause significant illness. Community members who interact with the ocean are exposed to dangerous levels of toxicants, such as acetone, asbestos, methanol, xylene, plasticizers, hormones, flame retardants, and other compounds (Allsing et al., 2022; García et al., 2020). In a study released in 2023, San Diego State University researchers identified 392 organic chemical contaminants in the Tijuana River and 175 of which are listed under the EPA Toxic Substance Control Act (Espinoza, 2023). These substances are heavily regulated by the EPA because they carry substantial risk of injury to human health or the environment. Along with a laundry list of hazardous chemicals, concerning bacterial and viral strains of *Mycobacterium tuberculosis, Salmonella, Vibrio, Streptococcus, HIV-1, and Hepatitis B and C* have been identified in the water. Furthermore, two bacteria strains of *E. coli* and *Legionella* have presented with antibiotic resistant genes making them impervious to penicillin (Allsing et al., 2022). These ocean bound contaminants such as bacteria, viruses, and industrial chemicals have been demonstrated to become aerosolized and suspended in the atmosphere through wave action in the surf zone (Pendergrath et al., 2023). Contaminant levels steadily increase in the air after rainfall ceases (Cooper et al., 2025). They are then blown 668 m or nearly half a mile inland (Pendergrath et al., 2021) by onshore winds, exposing thousands of people who never come into direct contact with the ocean.

These studies have primarily been conducted as reactionary one-off investigations by either government entities or local universities as opposed to routine preventative monitoring. This has led to large knowledge gaps about what is entering the ocean and what people are being exposed to. Furthermore, lack of monitoring has led to an erosion of community trust in the entities that are meant to protect them. Without consistent monitoring, there is essentially a blackhole of information because we can't know what's there if we don't test for it. Due to the health impacts, beaches in Imperial Beach were continuously closed for water activities for over 1300 days (Rico et al., 2025), forcing south San Diego surfers to leave their home break behind and seek waves further north. The water may be officially closed but that doesn't stop some diehard locals or ignorant tourists from testing out their luck.

The dangers of the ocean don't stop just because a little yellow sign on the beach says so. When people who enter the water inadvertently need help, first responders such as Imperial Beach, Coronado, California State Parks, and San Diego Lifeguards are the first in line to be put at risk. During critical rescue situations, lifeguards don't have the luxury to decide whether to enter the water. The very nature of water rescues in the ocean doesn't allow for much personal protective equipment (PPE). There's no way to keep water from entering the mouth while sprinting out through a rip current in overhead surf. Even if a lifeguard used a few precious seconds to put in ear plugs, throw on a dive mask and pull on a wetsuit to provide some level of protection, often, those are seconds a struggling victim doesn't have. Ask any ocean lifeguard and they'll tell you a least one story where if they had gotten to a victim even half a second later, they wouldn't have survived. Lifeguards provide a vital safety net for ocean goers by spending thousands of hours watching the water and anticipating the ocean's next move. Their entire job description requires them to think about the safety of others while putting themselves at risk when the time to perform a rescue inevitably comes. While the surf may be a toxic sludge fest, if lifeguards don't go, victims will succumb to the power of the ocean.

City of San Diego Lifeguards, who are the primary response for river rescues during rain events and flooding, have reported that their thick rubber boots and dry suits begin to melt after being exposed to pollutants in the Tijuana River. Lifeguard PPE begins to dissolve after only one wet season and significantly increases the rate of occupational exposure to contaminants. When lifeguards enter the water under these conditions, once returning to shore, many agencies have adopted a policy requiring guards to fill out an exposure report. These reports are utilized to document harmful exposure and to cover subsequent water contact related illnesses that arise from swimming through contaminated water. While I was never a member of the River Rescue Team as a seasonal lifeguard, I experienced firsthand filling out an exposure report due to dermal contact with runoff. I found myself performing an off-duty rescue in a flooded creek during a storm right around the corner from my apartment. The stormwater runoff flowing past myself, and the victim was a grayish color with a sewage odor, and an oily sheen coating the surface. Luckily, I never got sick afterwards, but I attribute that to the fact that I never dunked my head underwater. When members of highly specialized teams like the San Diego River Rescue team become ill due to occupational exposure and can't meet staffing needs, the situation evolves from a public health crisis to a public safety crisis as well.

Public health and safety crises like the one emanating from the Tijuana River will likely continue to increase in the future as climate change exacerbates rainfall events and flooding (Dettinger, 2011). Larger storms will continue to strain or even destroy critical infrastructure which will lead to an influx of pollutants entering oceans worldwide. As the effects of climate change increase, so will the technological advancements of humanity. New chemicals, products, and manufacturing techniques are constantly being developed. If control mechanisms and regulations are not updated to address these new challenges, these new advancements will also lead to the contamination of bodies of water (Xabadia et al., 2021). Without robust preparation for the future today, widespread pollution will only get worse.

Staying with the Trouble: Active and Resigned Activism in the Face of Polluted Spaces

In Donna Haraway's work, *Staying with the Trouble,* (Haraway, 2016, pg. 4) she emphasizes, we must make "oddkin"; we require each other in unexpected collaborations and combinations to navigate complex problems. In an increasingly interconnected world, we would be wise to heed her advice and form strong bonds between surfers and non-surfers alike to advocate for the ocean. While it might be difficult to fathom today, throughout the twentieth century the coastal waters off the Southern California Bight were littered with designated legal dumping areas for hazardous materials. Companies loaded up barges with radioactive waste from hospitals, discarded military munitions, chemical waste, petroleum waste, and toxic agrochemical insecticides like DDT, which were then dumped overboard (Schmidt et al., 2024). The millions of metric tons of waste dumped into the ocean over the decades remain on the ocean floor and continue to be released into sediment and the water column. Although DDT was banned in the United States in 1972, we continue to see it rear its ugly head through bioaccumulation in fishes and other marine wildlife throughout the Southern California region (McGill et al., 2024). DDT is classified as a man made persistent organic pollutant, meaning it's highly resistant to degradation through natural chemical, biological, or photolytic processes (Ritter et al., 2007). As with most pollutants, DDT trickles up the food chain until it is eventually consumed by humans without us even knowing it. Human exposure rates have fallen dramatically since DDT was banned over half a century ago, but it provides food for thought about the impacts of chemicals, plastics, and pesticides we use today on future generations.

The dump sites for hazardous materials were primarily located just offshore from harbors to provide easy access and to reduce fuel costs. There was a total of 14 permitted deep-water dumping sites off the coast of Ensenada, San Diego, Oceanside, Dana Point, Newport, Los Angeles and Ventura. A number of these sites utilized the same natural ocean bathymetry (underwater topography) that surfers crave to channel swell to create perfect waves. One of the sites primarily utilized by the U.S. Navy from 1945 to 1970 to dispose of explosives and toxic chemical ammunition was located at the mouth of the La Jolla submarine canyon leading to the world class break, Blacks. Furthermore, two sites that received 460,000 metric tons of sodium cyanide were situated only nine miles west of Sunset Cliffs (SCCRWRP, 1973). The information above only includes "legal" and reported waste dumped in the Southern California Bight. It's likely that there were millions more tons of hazardous and conventional waste illegally dumped in the ocean during that time period. Waste dumping in the ocean was officially outlawed in 1972 after the passage of the Ocean Dumping Act. The Ocean Dumping Act was one of many policies codified into law such as the Clean Water Act and Clean Air Act that came about through the galvanization of the environmental movement in the early 1970s. After decades of highly visible pollution dominating every aspect of American life, public outrage bubbled over into ardent and vocal activism. Hazardous waste dumping in coastal waters throughout the U.S.

provides a valuable case study demonstrating that positive change can happen, but only if the public demands it.

When it comes to tackling complex and widespread pollution sources, systemic change takes time and can be frustratingly slow. In an age of wanting everything yesterday, the idea of solutions taking decades to implement feels like watching paint dry. However, through dedicated funding, resources, and attention, human and environmental health can be substantially improved. Another positive example of this takes us to La Jolla in the early 2000's. Back when kegs and bonfires were legal on the beach and an iPhone didn't suddenly appear the second winds turned offshore, bacteria were also getting in on the party. Over 13 surfing beaches monitored for bacteria from Pacific Beach to La Jolla Shores were listed as having high bacteria levels in 2002. Through committed funding towards eliminating bacteria sources on land and stormwater infrastructure upgrades to divert low flows of runoff to the sewer system for treatment, the number of listed beaches has since dropped to just one in 2022. In the one site that remains, Children's Pool, the bacteria sources have been determined to be due to natural causes from fecal matter from harbor seals, sea lions, and birds that seek sanctuary in the protected cove. The process of finding and eliminating bacteria sources throughout La Jolla and upgrading infrastructure wasn't a linear one by any means. It took 20 years to finally get to the place we are today.

Substantial change requires time, especially when it comes to infrastructure upgrades and non-point source pollution abatement. Communities dealing with pollution often don't have the time to wait and need immediate relief to live their lives. When community members are left out in the cold by institutions meant to support them, citizen action takes shape as resigned activism. In South County San Diego, where the Tijuana River crisis permeates all aspects of life, community members have resorted to daily rituals of resigned activism. Residents close their windows at night, buy their own portable indoor air filters, don N95s or respirators to walk through their neighborhood and some have even gone as far as duct taping the seals around the windows and doors of their homes. A desire for a perceived sense of security in spaces meant to provide safety and relaxation crosses all demographics, cultures, and languages. In the wake of the humanitarian and environmental disaster from the Fukushima nuclear meltdown, surfers bought their own Geiger counters to take radiation measurements at their home beaches prior to paddling out for a session (Evers, 2021). Surfers need the sea to feel sane and will go to great lengths to glide along the face of a wave. Surfers and non-surfers alike in San Diego, Tijuana, and Fukushima are exposed to hazardous pollutants from the comfort of their own homes. This ever-present pollution separates communities from their sense of place, forcing community members to take extra precautions, often at their own financial detriment, to feel safe. Surfers around the world engage in a form of resigned activism every time they choose to not surf after a storm or wear ear plugs to prevent infection. While these actions benefit us individually, in the short term, surfing has managed to remain relatively pure and blind to the world's greatest social problems (Lueras, 1984). Collective action through active activism for clean water benefits not just surfers but the entire surrounding coastal community. Environmental justice encompasses not just a right to clean air and water but the ability to play in local blue spaces.

Future Implications: Emerging, Evolving, and Adaptive Technologies

Now that you've run the gauntlet of water quality information overload, some of you may think twice before paddling out after a storm. However, you may be like me and have all that information tucked away in the back of your mind, but the allure of scoring often outweighs health concerns. I personally choose the breaks I surf after a storm by identifying where the nearest storm drains are, how much surface area they drain, and how much rainfall was received. The larger the drainage area and the more rainfall that occurs all lead to higher pollution and bacteria levels. I generally try to avoid surfing the first day after a storm since the first 24 h are the most toxic. When I do paddle out during the 72 h window when bacteria levels are questionable at best, I take a few measures to reduce the risk of getting sick after the session. I use ear plugs to prevent both infection and reduce the risk of developing surfer's ear. I shower immediately after getting out of the water and scrub any open wounds with soap, or if I'm far from home, I'll use a gallon jug of water to rinse my eyes, ears, nose, and mouth right away. In the realm of health, a little prevention goes a long way. If all that nasty water is allowed to fester on and around your face, there's likely a higher chance of developing an infection down the road. Additionally, since I'm prone to sinus infections, I rinse my sinuses as soon as possible with a Neti Pot. Finally, I'll use a 1:1 ratio of hydrogen peroxide and warm water to dry out my ears, especially in the winter. I've found that all these little tricks help, but they don't always win against mother nature. I personally hate letting one session ruin a weeks' worth of sessions because of a sinus or respiratory infection.

Proactive and preventative measures can go a long way in ensuring safe, clean water for all. Many of the initiatives needed to make clean water a reality require dedicated and continual financial investment. Investments in critical sewer and stormwater infrastructure can reduce the volume of pollutants entering coastal waterways. However, infrastructure isn't always a glamorous or flashy use of tax dollars. Especially considering that water, sewer, and stormwater pipes are out of sight, out of mind. No one thinks twice after flushing the toilet until the 100 plus year-old sewer main fails, and raw sewage creates a chocolate river down the street. One of the first steps towards increasing funding towards these projects would be to express your support of water quality programs and policies to your elected representatives. We spend hours a day mindlessly scrolling on our little pocket machines. Instead of melting your brain, utilize some of that time to fire off an email or a phone call to your city council member, county supervisor, state representatives, and congressional representatives. Ask them to, at the bare minimum, support fully funding the BEACH Act, because after all, the squeaky wheel gets the grease. If the full $30 million allocated for the BEACH Act became a reality, your local county health department would receive three times the amount of funding it does now. The $20 million annual funding gap directly takes away from local resources that could be used to support other water quality initiatives such as infrastructure upgrades, special studies, pollution source tracking, and predictive modeling.

Times of crisis breeds innovation, whether it be through local emergencies like the Tijuana River crisis or the regional and global effects of climate change. Whether we like it or not, the future of our environment, societies, and surfing will continue to feel the compounding effects of a warming climate and extreme weather oscillation. Along the U.S. west coast, drastic swings between periods of intense rainfall followed by scorching droughts will become more frequent. These shifts will lead to less stormwater infiltration and larger pollutant loads being discharged into the ocean as well as, larger, more intense wildfires on land. Solutions to intricate and widespread issues require an overlapping mosaic of initiatives at the local, regional, and global level.

Stormwater has traditionally been viewed as a waste product to be removed from the built environment as quickly as possible. Stormwater in semi-arid Southern California is an increasingly valuable resource that shouldn't be wasted. High flow rates through channelized river bottoms and stream beds carry pollutants via a one-way express ticket to the ocean. Capture and reuse of stormwater through the implementation of blue-green infrastructure (BGI) not only increases green spaces within a community but also improves water quality. Stormwater capture bolsters groundwater aquifers that can be used for drinking water purposes and keeps surrounding vegetation greener, longer. Every gallon of water that can be captured onsite at residential, commercial or public spaces instead of being turned into runoff makes a difference. Less runoff directly correlates to fewer pollutants being suspended in the flow path and transported downstream.

The future of surfing as we continue into the Anthropocene will evolve just like every other aspect of our lives. Technological and scientific advancements are rapidly occurring in the environmental health, water quality, and surf modeling fields. The situation becomes a matter of being able to succinctly, equitably, and efficiently provide the information to the masses in an easily digestible format. Emerging technologies such as upstream real-time water quality monitoring sensors, rapid bacteria testing methods, and predictive pollution modeling will shape the future of surfing as we know it. Upstream sensors in storm drains and waterways calibrated to identify specific pollutants can send real time data to public health departments allowing them to issue advisories to communities prior to pollution entering the ocean. Even the most advanced bacteria testing methods like ddPCR have a delay between sampling and public notification of about 24 h. Finding out the surf break you paddled out at had high bacteria levels the day after you were getting tubed doesn't do you any favors. Increased and advanced monitoring directly increases public safety for beach communities. Lifeguards would be able to restart in-service trainings sooner when water quality returns to safe levels in order to maintain peak operational performance for rescue situations. It's one thing to watch a cranking rip current in double overhead surf from the tower. However, it's entirely different to feel the rip current moving through the sandbars or over the reef prior to making a critical rescue. Being in the ocean engages all the senses and expertly navigating the surf is just as much about the feel of the water as it is witnessing it.

The advancement and expansion of predictive modeling through the usage of nowcasts and forecasts will likely be the most revolutionary technology that will impact our surfing. A nowcast is a type of tool or system that uses computer models, runoff flow measurements, past bacteria data, watershed topography, tides, currents, mathematical techniques, satellite telemetry, and machine learning algorithms (AI) to provide estimates of water-quality threats in near-real-time. Predictive FIB nowcast models have been used for recreational water quality monitoring and beach management in the U.S. (Great Lake beaches in Michigan, Ohio, and New York) (Francy, 2009; Francy et al., 2013), the UK (Crowther et al., 2001), and Hong Kong (Thoe & Lee, 2014). Furthermore, the FIB nowcasts utilized within the Great Lakes have been shown to be as good as or better than grab sample methods (Francy et al., 2020). The development of these nowcasts are time and labor intensive due to the wide variety of site-specific variables that require past data spanning years needed to build an accurate model for each recreational beach (Searcy & Boehm, 2021). However, they are the most comprehensive and protective of overall health due to their ability to accurately display real time bacteria conditions. Some models have the capacity to provide forecasts as well, with some predicting bacteria levels three days in advance (Searcy & Boehm, 2022). In the case of the new Pathogen Forecast Model, developed by researchers at UC San Diego's Scripps Institution of Oceanography, sewage concentrations along the San Diego and Mexico coastline are capable of being predicted five days in advance (SIO, UCSD, 2025). The Pathogen Forecast Model provides an accurate and easy to understand risk assessment tool for surfers to plan their sessions in advance based on pollution levels along a thirty-mile stretch of coast. Similar technologies are also being deployed by the California-Harmful Algae Risk Mapping Model (C-HARM) to monitor for HABs. C-HARM has recently installed near, and offshore buoys equipped with underwater cameras that take photos every 25 min. The photos are then analyzed by AI to determine the likelihood of an HAB forming and the rate of spread along the coast. As these technologies progress, it will be imperative to ensure data collected by public agencies with public tax dollars isn't locked away behind a paywall. One of the cornerstones of public and environmental health is having access to reputable and free data. I foresee predictive bacteria modeling data along with real time water quality measurements being tied into a segment of Surfline and other surf forecasting models just like winds, tides, and swell direction. These technologies and the subsequent data will provide another tool in the risk vs. reward toolbox for surfers when deciding whether to paddle out.

At the end of the day, we all just want to feel safe and secure, especially in spaces that are meant to provide joy, relaxation, and leisure like our homes or the ocean. Future generations may view those of us who surf a river mouth during a storm or in front of a flowing storm drain as crazy due to the exposure to pollution we accept. Many of us will be able to look back on this time and tell our grandchildren about the chocolate-colored barrels we scored just after a storm that left our wetsuits smelling like gasoline, eyes watering, ears pounding, and skin itching for days afterwards. Their reactions will hopefully be that of disbelief and insist that grandma/grandpa took one too many on the head because there's no way that's how it used to be in the old days. The "good ol' days" of water quality at our local breaks are very likely still ahead of us. However, it will take collective and consistent action to advocate for safeguards to protect the ocean and our health as we immerse ourselves in the art of surfing.

References

Ackerman, D., & Weisberg, S. B. (2003). Relationship between rainfall and beach bacterial concentrations on Santa Monica Bay beaches. *Journal of Water and Health, 1*(2), 85–89.

Allsing, N., Kelley, S. T., Fox, A. N., & Sant, K. E. (2022). Metagenomic analysis of microbial contamination in the US portion of the Tijuana River watershed. *International Journal of Environmental Research and Public Health, 20*(1), 600.

Arnold, B. F., Schiff, K. C., Ercumen, A., Benjamin-Chung, J., Steele, J. A., Griffith, J. F., et al. (2017). Acute illness among surfers after exposure to seawater in dry-and wet-weather conditions. *American Journal of Epidemiology, 186*(7), 866–875.

Arnold, B. F., Wade, T. J., Benjamin-Chung, J., Schiff, K. C., Griffith, J. F., Dufour, A. P., et al. (2016). Acute gastroenteritis and recreational water: Highest burden among young US children. *American Journal of Public Health, 106*(9), 1690–1697.

Bertrand-Krajewski, J. L., Chebbo, G., & Saget, A. (1998). Distribution of pollutant mass vs volume in stormwater discharges and the first flush phenomenon. *Water Research, 32*(8), 2341–2356.

Bradley, G., & Hancock, C. (2003). Increased risk of non-seasonal and body immersion recreational marine bathers contacting indicator microorganisms of sewage pollution.

Brand, L. E., & Compton, A. (2007). Long-term increase in Karenia brevis abundance along the Southwest Florida coast. *Harmful Algae, 6*(2), 232–252.

Brand, L. E., Pablo, J., Compton, A., Hammerschlag, N., & Mash, D. C. (2010). Cyanobacterial blooms and the occurrence of the neurotoxin, beta-N-methylamino-l-alanine (BMAA), in South Florida aquatic food webs. *Harmful Algae, 9*(6), 620–635.

Britton, E., Kindermann, G., Domegan, C., & Carlin, C. (2020). Blue care: A systematic review of blue space interventions for health and wellbeing. *Health Promotion International, 35*(1), 50–69.

Buckley, R. (2002). Surf tourism and sustainable development in indo-Pacific Islands. I. The industry and the islands. *Journal of Sustainable Tourism, 10*(5), 405–424.

Buckley, R. C., & Cooper, M. A. (2023). Mental health contribution to economic value of surfing ecosystem services. *npj Ocean Sustainability, 2*(1), 20.

California Coastal Commission (2023, September 29). Tijuana river pollution crisis in San Diego county. W6b-10-2023-appendix.pdf.

US Census Bureau. (2021, April 26). Historical population change data (1910–2020). https://www.census.gov/data/tables/time-series/dec/popchange-data-text.html

Chelsea, N. R., Graeme Lockaby, B., Kalin, L., & Anderson, C. (2012). Effects of urbanization on stream hydrology and water quality: The Florida Gulf Coast. *Hydrological Processes, 26*(13), 2019–2030.

Colford, J. M., Jr., Wade, T. J., Schiff, K. C., Wright, C. C., Griffith, J. F., Sandhu, S. K., et al. (2007). Water quality indicators and the risk of illness at beaches with nonpoint sources of fecal contamination. *Epidemiology, 18*(1), 27–35.

Cooper, A., Cancelada, L., Torres, R. R., Belcher, K., Small, M., Belda-Ferre, P.,... & Prather, K. A. (2025). Identifying wastewater chemicals in coastal aerosols. *Science Advances, 11*(22), eads9476.

Crowther, J., Kay, D., & Wyer, M. D. (2001). Relationships between microbial water quality and environmental conditions in coastal recreational waters: The Fylde coast. *UK. Water research, 35*(17), 4029–4038.

Danil, K., Colegrove, K. M., Delaney, M. A., Mena, A., Stedman, N., & Wurster, E. (2023). Systemic erysipelas outbreak among free-ranging bottlenose dolphins, San Diego, California, USA, 2022. *Emerging Infectious Diseases, 29*(12), 2561.

DeFlorio-Barker, S., Wing, C., Jones, R. M., & Dorevitch, S. (2018). Estimate of incidence and cost of recreational waterborne illness on United States surface waters. *Environmental Health, 17*, 1–10.

Department of Environmental Health and Quality (DEHQ). (2024). General rain advisory for all coastal waters. County of San Diego. https://www.sandiegocounty.gov/content/dam/sdc/deh/lwqd/Beach&Bay/DEHQ_bb_public_summary.pdf

Dettinger, M. (2011). Climate change, atmospheric rivers, and floods in California–a multimodel analysis of storm frequency and magnitude changes 1. *JAWRA Journal of the American Water Resources Association, 47*(3), 514–523.

Dias, M. (2024, February 21). *Beach act of 2024 aims to improve public health protection at U.S. beaches*. Surfrider Foundation.

Dwight, R. H., Baker, D. B., Semenza, J. C., & Olson, B. H. (2004). Health effects associated with recreational coastal water use: Urban versus rural California. *American Journal of Public Health, 94*(4), 565–567.

Dwight, R. H., Semenza, J. C., Baker, D. B., & Olson, B. H. (2002). Association of urban runoff with coastal water quality in Orange County, California. *Water Environment Research, 74*(1), 82–90.

Environmental Protection Agency (2024, October 15). Beach Management Resources. *EPA*. https://www.epa.gov/beaches/about-beach-act-The-BEACH-Act-Grant-Program-beach-monitoring%20-and%20notification%20programs

Environmental Protection Agency (2024, July). Office of water 823-R-24-002. *EPA's Beach Report: 2023 Swimming Season.* https://nepis.epa.gov/Exe/ZyPDF.cgi/P101B1MA.PDF?Dockey=P101B1MA.PDF

Environmental Protection Agency (2025, September). USMCA Tijuana River Watershed EPA.gov. https://www.epa.gov/sustainable-water-infrastructure/usmca-tijuana-river-watershed#transboundary. Retrieved October 27, 2025.

Espinoza, J. (2023). *Identification and assessment of constituents of emerging concern in the Tijuana River Stormwater*. (Master's thesis, San Diego State University.

Evers, C. (2021). Polluted leisure and blue spaces: More-than-human concerns in Fukushima. *Journal of Sport and Social Issues, 45*(2), 179–195.

Feddersen, F., Boehm, A. B., Giddings, S. N., Wu, X., & Liden, D. (2021). Modeling untreated wastewater evolution and swimmer illness for four wastewater infrastructure scenarios in the San Diego-Tijuana (US/MX) border region. *GeoHealth, 5*(11), e2021GH000490.

Fleisher, J. M., Fleming, L. E., Solo-Gabriele, H. M., Kish, J. K., Sinigalliano, C. D., Plano, L., et al. (2010). The BEACHES study: Health effects and exposures from non-point source microbial contaminants in subtropical recreational marine waters. *International Journal of Epidemiology, 39*(5), 1291–1298.

Francy, D. S. (2009). Use of predictive models and rapid methods to nowcast bacteria levels at coastal beaches. *Aquatic Ecosystem Health & Management, 12*(2), 177–182.

Francy, D. S., Brady, A. M., Cicale, J. R., Dalby, H. D., & Stelzer, E. A. (2020). Nowcasting methods for determining microbiological water quality at recreational beaches and drinking-water source waters. *Journal of Microbiological Methods, 175*, 105970.

Francy, D. S., Stelzer, E. A., Duris, J. W., Brady, A. M., Harrison, J. H., Johnson, H. E., & Ware, M. W. (2013). Predictive models for Escherichia coli concentrations at inland lake beaches and relationship of model variables to pathogen detection. *Applied and Environmental Microbiology, 79*(5), 1676–1688.

Freeman, L. A., Corbett, D. R., Fitzgerald, A. M., Lemley, D. A., Quigg, A., & Steppe, C. N. (2019). Impacts of urbanization and development on estuarine ecosystems and water quality. *Estuaries and Coasts, 42*, 1821–1838.

Fujioka, R. S., Hashimoto, H. H., Siwak, E. B., & Young, R. H. (1981). Effect of sunlight on survival of indicator bacteria in seawater. *Applied and Environmental Microbiology, 41*(3), 690–696.

Gajski, G., Kašuba, V., Milić, M., Gerić, M., Matković, K., Delić, L., et al. (2024). Exploring cytokinesis block micronucleus assay in Croatia: A journey through the past, present, and future in biomonitoring of the general population. *Mutation Research-Genetic Toxicology and Environmental Mutagenesis, 895*, 503749.

García, J., García-Galán, M. J., Day, J. W., Boopathy, R., White, J. R., Wallace, S., & Hunter, R. G. (2020). A review of emerging organic contaminants (EOCs), antibiotic resistant bacteria (ARB), and antibiotic resistance genes (ARGs) in the environment: Increasing removal with wetlands and reducing environmental impacts. *Bioresource Technology, 307*(123228), 7.

Gibbs, K., Wilkie, L., Jarman, J., Barker-Smith, A., Kemp, A. H., & Fisher, Z. (2022). Riding the wave into wellbeing: A qualitative evaluation of surf therapy for individuals living with acquired brain injury. *PLoS One, 17*(4), e0266388.

Given, S., Pendleton, L. H., & Boehm, A. B. (2006). Regional public health cost estimates of contaminated coastal waters: A case study of gastroenteritis at Southern California beaches.

Gobler, C. J. (2020). Climate change and harmful algal blooms: Insights and perspective. *Harmful Algae, 91*, 101731.

Gurtner, Y. (2004). After the Bali bombing—The long raod to recoverty. *The Australian Journal of Emergency Management, 19*, 56–66.

Haile, R. W., Witte, J. S., Gold, M., Cressey, R., McGee, C., Millikan, R. C., et al. (1999). The health effects of swimming in ocean water contaminated by storm drain runoff. *Epidemiology, 10*(4), 355–363.

Haraway, D. J. (2016). *Staying with the trouble*. Duke University Press.

Harding, A. K., Stone, D. L., Cardenas, A., & Lesser, V. (2015). Risk behaviors and self-reported illnesses among Pacific Northwest surfers. *Journal of Water and Health, 13*(1), 230–242.

Honkanen, I. (2020). Environmental Health and Policy at the Tijuana-San Diego Border. *Undergraduate Honors Theses.*, 72.

Johnsen, C. (2018). Countries divided, shared pollution: Whose mess is it?. A Case Study on the issue of Transnational Border Sewage in the Tijuana River Valley Watershed.

Katz, C., Rosen, G., & Arias, E. (2006). Storm water toxicity evaluation conducted at Naval Station San Diego, naval Submarine Base San Diego, naval Amphibious Base Coronado, and Naval Air Station North Island. In *SPAWAR Systems Center Pacific Technical Report 1938*.

Krawczyk, B., Wityk, P., Gałęcka, M., & Michalik, M. (2021). The many faces of enterococcus spp.—Commensal, probiotic and opportunistic pathogen. *Microorganisms, 9*(9), 1900.

Laderman, S. (2014). *Empire in waves: A political history of surfing*. University of California Press: Berkeley.

Leecaster, M. K., & Weisberg, S. B. (2001). Effect of sampling frequency on shoreline microbiology assessments. *Marine pollution bulletin, 42*(11), 1150–1154.

Leonard, A. F., Singer, A., Ukoumunne, O. C., Gaze, W. H., & Garside, R. (2018). Is it safe to go back into the water? A systematic review and meta-analysis of the risk of acquiring infections from recreational exposure to seawater. *International Journal of Epidemiology, 47*(2), 572–586.

Li, J., & Zhang, X. (2019, Jun 4). Beach pollution effects on health and productivity in California. *International Journal of Environmental Research and Public Health, 16*(11), 1987. PMID: 31167463; PMCID: PMC6604031. https://doi.org/10.3390/ijerph16111987

Lueras, L. (1984). *Surfing: The ultimate pleasure*. New York: Workman Publishing.

Mach, L., Rothrock, E., Stark, S., & Nahmias, J. (2024). Place attachment, wellbeing, and conservation in surf destination communities. *Frontiers in Sustainable Tourism, 3*, 1387081.

Maquiladoras/Twin Plants (2025). Maquiladoras/twin plants city of San Diego official website. https://www.sandiego.gov/economicdevelopment/sandiego/trade/mexico/maquiladoras#:~:text=burdened%20hourly%20basis.-,Maquiladoras'%20Effect%20on%20San%20Diego,importance%20as%20a%20manufacturing%20area

McFarland, A. R., Larsen, L., Yeshitela, K., Engida, A. N., & Love, N. G. (2019). Guide for using green infrastructure in urban environments for stormwater management. *Environmental Science: Water Research & Technology, 5*(4), 643–659.

McGill, L., Sleugh, T., Petrik, C., Schiff, K., McLaughlin, K., Aluwihare, L., & Semmens, B. (2024 Nov 5). The persistent DDT footprint of ocean disposal, and ecological controls on bioaccumulation in fishes. *Proceedings of the National Academy of Sciences of the United States of America, 121*(45), e2401500121. https://doi.org/10.1073/pnas.2401500121. Epub 2024 Oct 28.

Moore, S. K., Trainer, V. L., Mantua, N. J., Parker, M. S., Laws, E. A., Backer, L. C., & Fleming, L. E. (2008, November). Impacts of climate variability and future climate change on harmful algal blooms and human health. In *Environmental health* (Vol. 7, pp. 1–12).

Noble, R.T., Blackwood, A.D., Griffith, J.F., McGee, C.D. & Weisberg, S.B. (2010, November). Comparison of rapid quantitative PCR-based and conventional culture-based methods for enumeration of Enterococcus spp. and Escherichia coli in recreational waters. *Applied and Environmental Microbiology. 76*(22), 7437–7443. Epub 2010 Sep 24. PMID: 20870786; PMCID: PMC2976187. https://doi.org/10.1128/AEM.00651-10

O'Halloran, C., & Silver, M. (2021). Health risks and benefits among surfers after exposure to seawater in Monterey bay, Santa Cruz county, California, United States. *Frontiers in Marine Science, 8*, 714831.

OFFICE OF WATER 820-F-058 (2012). *Recreational Water Quality Criteria.* https://www.epa.gov/sites/default/files/2015-10/documents/rwqc2012.pdf

O'Halloran, C. L. (2011). *Oceans and human health: Harmful algal blooms and acute health symptoms among surfers and lifeguards.* University of California.

Old, J. (2020, February). Transnational surfistas: Surf transnationalism in Nicaragua's Emerald Coast. In *Proceedings of the 6th Biennial Global Wave Conference, Gold Coast, Australia*, pp. 10–14.

Olive, L., Dober, M., Mazza, C., Turner, A., Mohebbi, M., Berk, M., & Telford, R. (2023). Surf therapy for improving child and adolescent mental health: A pilot randomised control trial. *Psychology of Sport and Exercise, 65*, 102349.

Olive, R., & Wheaton, B. (2021). Understanding blue spaces: Sport, bodies, wellbeing, and the sea. *Journal of Sport and Social Issues, 45*(1), 3–19.

Pendergraft, M. A., Belda-Ferre, P., Petras, D., Morris, C. K., Mitts, B. A., Aron, A. T., et al. (2023). Bacterial and chemical evidence of coastal water pollution from the Tijuana river in sea spray aerosol. *Environmental Science & Technology, 57*(10), 4071–4081.

Pendergraft, M. A., Grimes, D. J., Giddings, S. N., Feddersen, F., Beall, C. M., Lee, C., et al. (2021). Airborne transmission pathway for coastal water pollution. *PeerJ, 9*, e11358.

Ponting, J. (2009). Projecting paradise: The surf media and the hermeneutic circle in surfing tourism. *Tourism Analysis, 14*(2), 175–185.

Rexin, D., Rachmadi, A.T. & Hewitt, J. (2024, March). Persistence of infectious human norovirus in estuarine water. *Food and Environmental Virology. 16*(1):58–64. Epub 2024 Jan 2. PMID: 38165609; PMCID: PMC10963509. https://doi.org/10.1007/s12560-023-09577-w

Rico, B., Barsanti, K. C., Porter, W. C., Cysneiros de Carvalho, K., Stigler-Granados, P., & Prather, K. A. (2025). Heavily polluted Tijuana River drives regional air quality crisis. *Science, 389*(6763), eadv1343.

Ritter, L., Solomon, K.R., Forget, J., Stemeroff, M. & O'Leary, C. (2007) Persistent organic pollutants. *United Nations Environment Programme.* Archived from (PDF) on 2007-09-26. Retrieved 16 September 2007.

Robinson, L., & Jarvie, J. (2008). Post-disaster community tourism recovery: The tsunami and Arugam Bay. *Sri Lanka. Disasters, 32*, 631–645.

Rozen, Y., & Belkin, S. (2001). Survival of enteric bacteria in seawater. *FEMS Microbiology Reviews, 25*(5), 513–529.

SCCWRP Staff. (1973). *The ecology of the Southern California Bight: Implications for water quality management*. Technical Report 10. Southern California Coastal Water Research Project. El Segundo.

Schiff, K. C., Griffith, J. F., Steele, J. A., Arnold, B., Ercumen, A., Benjamin-Chung, J., Colford, J. M., Jr., Soller, J., Wilson, R., & McGee, C. (2016). The Surfer Health Study: A Three-Year Study Examining Illness Rates Associated with Surfing During Wet Weather. In *Technical report 943*. Costa Mesa, CA.

Schmidt, J. T., Wu, M. S. C., Kittner, H. E., Arey, J. S., Hammond, D. E., Earth 182A Group, & Valentine, D. L. (2024). Disentangling the history of Deep Ocean disposal for DDT and other industrial waste off Southern California. *Environmental Science & Technology, 58*(9), 4346–4356.

Schueler, T. R. (1994). Watershed protection techniques: A quarterly bulletin on urban watershed restoration and protection tools. *Center for Water Protection, 1*(2), 39–91.

Scott, S. Q., & Rogers, S. H. (2018). Surf's up? How does water quality risk impact surfer decisions? *Ocean & Coastal Management, 151*, 53–60.

Scripps Institution of Oceanography, University of California, San Diego. (2025) Pathogen Forecast Model. https://pfmweb.ucsd.edu. Accessed October 3, 2025.

Searcy, R. T., & Boehm, A. B. (2021). A day at the beach: Enabling coastal water quality prediction with high-frequency sampling and data-driven models. *Environmental Science & Technology, 55*(3), 1908–1918.

Searcy, R. T., & Boehm, A. B. (2022). Know before you go: Data-driven beach water quality forecasting. *Environmental Science & Technology, 57*(46), 17930–17939.

Steele, J. A., Blackwood, A. D., Griffith, J. F., Noble, R. T., & Schiff, K. C. (2018). Quantification of pathogens and markers of fecal contamination during storm events along popular surfing beaches in San Diego, California. *Water Research, 136*, 137–149.

Stigler Granados, P., Sant, K., Quintana, P. Hoh, E. Oren, E., Lopez-Galvez, N., Zavala Perez, M., & NI, Y. (2024). *Tijuana River Contamination from Urban Runoff and Sewage: A Public Health Crisis at the Border.* https://www.sdsu.edu/_files/tijuana-sewage-contamination-public-health-crisis-white-paper-021424.pdf

Strokal, M., Bai, Z., Franssen, W., Hofstra, N., Koelmans, A. A., Ludwig, F., et al. (2021). Urbanization: An increasing source of multiple pollutants to rivers in the 21st century. *npj Urban sustainability, 1*(1), 1–13.

Surfrider Foundation (2018, May 17). Beaches environmental assessment and coastal health (beach) Act of 2000. *Beachapedia.* https://beachapedia.org/(BEACH)_Act_of_2000#:.7E:text.3DThe_BEACH_Act_was_authorized.2Cfunding_level_to_.2460_millionBeaches_Environmental_Assessment_and_Coastal_Health.

Taylor, K. (2016, July 20). Contagion present. *Surfer Magazine.* http://www.surfermag.com/features/contagion-present. Accessed September 20, 2025.

Thoe, W., & Lee, J. H. (2014). Daily forecasting of Hong Kong beach water quality by multiple linear regression models. *Journal of Environmental Engineering, 140*(2), 04013007.

Tseng, L. Y.,Ts & Jiang, S. C. (2012). Comparison of recreational health risks associated with surfing and swimming in dry weather and post-storm conditions at Southern California beaches using quantitative microbial risk assessment (QMRA). *Marine Pollution Bulletin, 64*(5), 912–918.

Wade, T. J., Calderon, R. L., Sams, E., Beach, M., Brenner, K. P., Williams, A. H., & Dufour, A. P. (2006). Rapidly measured indicators of recreational water quality are predictive of swimming-associated gastrointestinal illness. *Environmental Health Perspectives, 114*(1), 24–28.

Wavelength Surfing in the time of Ebola (2016, 18 July). *Wavelength Magazine.* Available online: https://wavelengthmag.com/surfing-in-the-time-of-ebola/

White, M., Smith, A., Humphryes, K., Pahl, S., Snelling, D., & Depledge, M. (2010). Blue space: The importance of water for preference, affect, and restorativeness ratings of natural and built scenes. *Journal of Environmental Psychology, 30*(4), 482-493..

White, M. P., Elliott, L. R., Gascon, M., Roberts, B., & Fleming, L. E. (2020). Blue space, health and Well-being: A narrative overview and synthesis of potential benefits. *Environmental Research, 191*, 110169.
World Health Organization. (1946). *Constitution of the World Health Organization*. World Health Organization. https://www.who.int/about/governance/constitution
Wright, A. (1985). *Bank Surfing California, ©1973*. Mountain & Sea Publishing, p. 172
Xabadia, A., Ortuzar, I., Martinez, Y., & Esteban, E. (2021). Contaminants of emerging concern: A review of biological and economic principles to guide water management policies. *International review of environmental and resource economics*, (ART-2021-125430).
Xu, W., & Zhang, Z. (2022). Impact of coastal urbanization on marine pollution: Evidence from China. *International Journal of Environmental Research and Public Health, 19*(17), 10718.
Yabsley, N. A., Gilby, B. L., Schlacher, T. A., Henderson, C. J., Connolly, R. M., Maxwell, P. S., & Olds, A. D. (2020). Landscape context and nutrients modify the effects of coastal urbanisation. *Marine Environmental Research, 158*, 104936.
Yau, V., Wade, T. J., de Wilde, C. K., & Colford, J. M. (2009). Skin-related symptoms following exposure to recreational water: A systematic review and meta-analysis. *Water Quality, Exposure and Health, 1*, 79–103.
Yau, V. M., Schiff, K. C., Arnold, B. F., Griffith, J. F., Gruber, J. S., Wright, C. C., et al. (2014). Effect of submarine groundwater discharge on bacterial indicators and swimmer health at Avalon Beach, CA, USA. *Water Research, 59*, 23–36.

Part III
Society

Chapter 10
From Soul to Science: The Evolution of Psychedelics and the Future of Health, Healing, and Wellbeing in Surf Culture

Jarrett Robert Rose

Early Daze: The Brotherhood of Eternal Love, Timothy Leary, and the Soul of Soul Surfing

One of the most iconic and influential periods in surf history emerged from the subculture known as "soul surfing." Beginning in the 1950s, surfing began to grow rapidly, thanks in large part to surf films, which depicted the sport not only as unique and challenging, but as a far-out and radical subculture existing on the fringe of society (Booth, 1994, 1996). In many ways, these themes prepared the culture to merge seamlessly with the growing countercultural ethos of the 1960s, defined by one of its many mantras: "sex, drugs, and rock 'n' roll." As the two anti-establishment groups melted together, surfing entered a distinctive and enduring phase, one that marked the convergence of sport, art, and countercultural ideals, blended hedonistic exploration with deeper philosophical ideas, and created a lasting influence on the identity of modern surfing that can still be witnessed today (Lawler, 2011).

On the one hand, the tradition known as soul surfing can be described as part rejection of the "modernization" of surfing, consisting of the shortboard revolution and the concomitant growth in competitiveness and monetization. Smaller boards meant more power, intensity, and aggression, with sharper rails that enabled deeper cuts and faster turns—a far cry from its predecessor, the longboard, which manifested elegance, style, and stability. Yet on the other hand—and more important for the purposes of this chapter—soul surfing took on a life of its own, evolving into a distinct style of living. As Booth argues, it became an "amalgam of alternate, typically Utopian, lifestyles and political activism... an oppositional cultural practice symbolizing countercultural idealism" (1995: 195). At the center of this cultural turn was the use of drugs—most specifically, psychedelics like LSD, "magic mushrooms,"

J. R. Rose (✉)
Community & Behavioral Health Program and the Department of Sociology, State University of New York (SUNY) Polytechnic Institute, Utica, NY, USA
e-mail: rosejr@sunypoly.edu

L. Mach (ed.), *The Future of Surfing in the Anthropocene*,
https://doi.org/10.1007/978-3-032-09377-6_10

and mescaline. And one group in particular played an outsize role in inscribing the *soul* into the soul surfing movement: the Brotherhood of Eternal Love.

Founded in 1965 in Orange County, California, the Brotherhood of Eternal Love began as a group of outlaw hippies, surfers, and car enthusiasts. Yet, over time, they evolved into what was likely one of the nation's largest drug-smuggling networks, employing creative methods like hollowed-out surfboards to transport marijuana—and later LSD—around the globe. With their own brand of acid, Orange Sunshine, the Brotherhood not only shaped surf culture but also supplied their product to figures and groups spanning the counterculture, from the Grateful Dead, The Doors, and Jefferson Airplane to Charles Manson, the Hells Angels, and major festivals like Altamont, Newport Pop, and Woodstock. The Brotherhood held significant social and cultural capital—drawing in influential surfers like Mike Hynson (*The Endless Summer*), David Nuuhiwa (1968 international surf champion), Gerry Lopez ("Mr. Pipeline"), and Nat Young (Australian surf champion), as well as larger-than-life figures like Jimi Hendrix, whose last official concert in 1970 was in Hawaii with the Brotherhood while filming the infamous *Rainbow Bridge*. Their orbit profoundly shaped the era's surf culture and extended far beyond it.

Yet, the Brotherhood were more than just countercultural fugitives and drug smugglers with a global reach. At the heart of their drug-infused philosophy was one of radical spiritual transcendence. One man helped bring this specific ethos to life within their collective: the notorious, enigmatic Timothy Leary. Once a Harvard psychologist and later a defining countercultural figure of the 1960s and '70s, Leary arguably went further than anyone else of his era in experiencing, studying, and promoting psychedelic consciousness. More than just an advocate, he played a central role in *shaping* how psychedelics were understood and used. His influence stemmed largely from his emphasis on "set and setting," a now widely recognized principle in guiding psychedelic experiences.

Psychedelics are a strange class of psychoactive substances. While some individuals endure experiences that border on temporary insanity or psychosis—commonly referred to as "bad trips" or "challenging experiences"—others embark on extraordinary journeys that can have remarkable spiritual and therapeutic benefits (as indigenous peoples have known for centuries and Western science is finally learning). The reason for this is that psychedelics are heavily influenced by the "set and setting" in which they are taken—the personal *mindset* of the user and the sociocultural *setting* they are ingested in. This inherently psychological, social, and cultural context—set and setting—is the most crucial variable influencing the psychedelic experience. It helps explain both the historical and contemporary diversity of experiences people have with these substances. Timothy Leary not only explored the far reaches of the mind but also studied and wrote extensively about the intricate role of set and setting in shaping psychedelic experiences. It was this knowledge that, in the early 1960s, he transmitted to surf culture through the Brotherhood.

As Leary writes in his memoir, one night at a party in the Hollywood hills, he "met a delegation of athletic blonde men, surfers from Laguna Beach" (Leary 1997: 135). By the time of this chance encounter, the group was already well-acquainted with Leary's work and character. They had studied his books and academic papers, and, notably, embraced his countercultural call to 'turn on' newcomers to the psychedelic experience, spreading the psychedelic gospel far and wide. I have previously written about this fascinating time in Southern California and the influence it had on surf culture (Rose, 2024b). But what's important to recognize in this history of surfing's encounter with early psychedelia is the way in which the specific cultural context influencing surfers' psychedelic experience was driven by the philosophical synthesis of Leary and the Brotherhood.

The Spiritual and Revolutionary Core of Soul Surfing

Before we can fully understand the contemporary landscape of psychedelic experiences in surf culture, it's essential to examine the journey that brought us here. In other words, it's important to understand how the psychedelic experience was originally described and conceptualized. For Leary, psychedelics were largely understood through the work and framing of Aldous Huxley, who explained his own experience on mescaline (in *Doors of Perception*) as one of personal and spiritual transcendence. Later, in *The Psychedelic Experience*—coauthored by Leary, Richard Alpert (later known as Ram Dass), and Ralph Metzner, and based on *The Tibetan Book of the Dead*, which served as a key guide to psychedelic consciousness for the counterculture—it was argued that under the right set and setting, psychedelics could offer a "journey to new realms of consciousness" and facilitate "mystical" or "transcendental experiences." They described in nuanced detail the entire conscious experience through the framework of Classical Hinduism, arguing that the psychedelic journey was wholeheartedly a spiritual one. Honing the set and setting appropriately, they argued, consciousness could be a tool directed to accomplish:

"(1)increased personal power, intellectual understanding, sharpened insight into self and culture, improvement of life situation, accelerated learning, professional growth"; (2) for duty, help of others, providing care, rehabilitation, rebirth for fellow man; (3) for fun, sensuous enjoyment, aesthetic pleasure, interpersonal closeness, pure experience; (4) for transcendence, liberation from ego and space-time limits; attainment of mystical union." (Leary et al., 1969 [1964] 99).

For Leary and advocates, it was this psychedelic worldview that would inspire a movement toward complete cultural revolution (with undeniably political implications). The core insights derived from psychedelic experiences, he and his followers began to believe, were not just meant to explore and challenge one's private beliefs about the self or to attain spiritual transcendence (and neither were they simply described as tools for mental health, as will be discussed below), but to contest existing social institutions and advocate for systemic change. For Leary, the spirituality that he argued was a component of the psychedelic experience should mix with politics. The ultimate goal, he argued, was to create a society where everyone could have the "personal freedom... to control... the technology that reaches your brain, whether through drugs, the telephone, or the television" (Leary, 1968: 10).

Thus, in speeches across the country, books, and articles written on the psychedelic experience, it was this "spiritual conceptualization of psychedelics as devotional sacraments," steered toward cultural revolution, that "informed much of the popular interest in the drugs during the 1960s" (Hartogsohn, 2022: 584). In fact, this is exactly what Leary meant in 1967 when he told tens of thousands of onlookers at the Human Be-In in San Francisco to "turn on [the mind to higher states of awareness], tune in [to the true self], and drop out [of conventional, materialistic, oppressive social structures]."

At the center of Leary's call to transform society was the Brotherhood (and, as a proxy, surf culture), who adhered fervently to Leary's directives. Every week, for example, they held "Sunday School" sessions, where they'd instruct between 50 and 200 participants on how to take LSD (Leary 1997: 136). As I have written previously, several of those newcomers were high-profile soul surfers like Hynson, Nuuhiwa, Lopez, and Young. And through these surfers—along with the thousands who were "turned on" to the spiritual and revolutionary aspects of psychedelics, and the many more who took Orange Sunshine in the 1960s and 70s—the Brotherhood became a powerful cultural force. Their legacy not only helped define the soul surfing tradition for generations but also contributed to broader societal shifts in how psychedelics were perceived and experienced during the era. As Young once stated about the times:

> Soul surfing became the fashion. The longhaired, back-to-nature surf stars of the time were heavily involved in meditation, astrology, health foods, the self, the striving for physical perfection... and drugs. Taking psychedelic drugs and smoking marijuana became the norm. Surfers began dropping out and experimenting with communal living... it was the "look inward" era of flower power, mysticism, the Maharishi and the popularization of Eastern cults, which affected everyone from the Beatles and the Rolling stones to surfers (Young, 1983: 110; from Lawler, 2011: 138-39)

It was only a matter of time before Leary's unorthodox and provocative promotion of psychedelics as catalysts for spiritual and societal transformation alarmed enough of the authorities. His charismatic yet polarizing public persona, combined with a string of headline-making legal battles and imprisonment on drug charges, only reinforced the association between psychedelics and countercultural defiance. By the early 1970s, the psychedelic wing of the counterculture received substantial backlash, culminating in the Controlled Substances Act of 1970, which classified psychedelics as Schedule I substances, effectively criminalizing their use and stalling legitimate scientific research for decades. To this day, many blame Leary for what they see as a reckless, theatrical, and (ironically) egocentric approach, arguing that his antics accelerated government crackdown and set back the progress of psychedelic science. Whereas the soul tradition lived on in both above- and belowground settings—namely through the aesthetics and lifestyle portrayed in surf films, magazines, and documentaries (see Lawler, 2011)—the radical and revolutionary aspects of the culture diminished over time.

The Renaissance: Medicalizing the Psychedelic Experience

After a decades-long hiatus in research following their prohibition, psychedelics have reemerged in mainstream science and popular culture. However, unlike the 1960s, when psychedelia was deeply intertwined with spiritual awakening and countercultural revolution, today's movement is largely shaped by scientific and medical narratives emphasizing the *therapeutic* potential of these substances. Framed, therefore, as *psychedelic-assisted therapy* (PAT), media representation predominantly focuses on a growing body of research suggesting that, when combined with psychotherapeutic modalities, the psychedelic experience can be directed towards alleviating (if not, as some have argued, *curing*) symptoms of various mental health conditions. These include, for example, substance abuse (alcohol and tobacco) (Garcia-Romeu et al., 2014; Johnson et al., 2014, 2017; Bogenschutz et al., 2018), obsessive-compulsive disorder (Moreno et al., 2006), cancer-related anxiety and depression (Ross et al., 2016; Griffiths et al., 2016), and treatment-resistant depression (Carhart-Harris et al., 2016; Roseman et al., 2018; Watts et al., 2017).

In fact, today the tide has shifted so far from the countercultural perspective that sociologist Danielle Giffort (2020) argues convincingly in *Acid Revival: The Psychedelic Renaissance and the Quest for Medical Legitimacy*, that the primary motivating force of today's psychedelic agenda is to move *away from* the damage Leary caused a generation ago. This involves stripping these substances of their revolutionary associations and rebranding them with a more mainstream image for public marketing. Neitzke-Spruill (2023) has summarized this moment succinctly,

> Owing to the pursuit of "respectability" and integration into contemporary biomedicine, the psychedelic science movement promises an apolitical revolution that will minimally disrupt existing social forms. Using a logic of neuronal transformation, psychedelics are posed as neurotechnologies that are consonant with middle class life and have utility as both transdiagnostic treatments for medicalized mental health disorders and tools for self-improvement in otherwise healthy people (Neitzke-Spruill, 2023: 171-72).

It's a fascinating time: once vilified, banned, ridiculed, and central to the early hysteria of the War on Drugs, today a wide range of people and professionals from academic researchers to mainstream journalists (and even microdosing parents - see Ayelet Waldman's *A Really Good Day*) are highlighting the profound medicinal benefits these drugs may offer. The journey these substances have undertaken has been far out, to say the least.

The Psychedelic Seascape of Today: The Role of Discourse in Shaping Culture

The question remains: To what extent has the role played by psychedelics in surfing changed in the twenty-first century? Does the spirit of *soul*, as surfing's quintessential pastime, still exist within the art, sport, and culture of surfing? Or, has it been replaced by a newer, *medicalized and scientific* philosophy? Have we left the days of Leary and the Brotherhood behind, embracing instead a safer, more responsible approach to psychedelics, as seen through the lens of clinical trials and psychiatric medicine? In other words, in these early days of the psychedelic renaissance, are we witnessing a substitution of the old for the new, the soul for the science, the spirit for the emotions? *And what might this say about the future of psychedelics in surfing?*

I set out to explore this question myself, gathering data from a number of sources, including podcast and media interviews, documentary films, social media posts, and articles written by or about surfers. In the following section, I'll highlight several influential figures who have openly discussed their psychedelic use in public forums. Of course, many still choose to remain silent, as stigma persists. Yet, as we'll see, that silence is giving way to a broader, more open, less stigmatized conversation.

A brief word on theory: I use the term "discourse," drawing from philosopher and historian Michel Foucault, to refer to the way certain ideas and ways of thinking become dominant in society. Discourse is shaped by power structures, cultural norms, and social institutions, and in turn, it reinforces those same structures by influencing what is considered knowledge, truth, and acceptable behavior. For example, as I have shown above, in the twenty-first century, psychedelics have been reconceptualized through the fields of science and medicine—powerful institutions that have, in the attempt to legitimize these drugs, redefined them as therapeutic tools for health and well-being rather than as countercultural or recreational substances. In a similar fashion, in this section I analyze how professional surfers—powerful icons in the sport and culture of surfing—talk about psychedelics, including their experiences, philosophies, and practices. I view their narratives as both *shaped by* contemporary medico-scientific discourse (the power and knowledge disseminated through scientific institutions, particularly those emerging from psychiatry, neuroscience, and psychedelic-assisted therapy) and, at the same time, evolving alongside the historico-cultural discourse of the soul surfing era (originating with the Brotherhood and Leary), which already has its own historical understanding of psychedelic consciousness and its effects. Today's surfers, as we will find, often bridge these past meanings with contemporary therapeutic understandings, creating a hybrid discourse that blends scientific authority with cultural memory and lifestyle aesthetics from a past era. In this way, the narratives of surfers using psychedelics today are both shaped by institutional power-knowledge and generative of new meanings within surf culture—offering models, metaphors, and references for others to watch, hear, follow, and emulate.

Kelly Slater, Jamie Mitchell, and Ayahuasca

In 2017, the winningest surfer in history travelled to a "medically licensed plant medicine center" in Costa Rica. Contemplating the move for quite a while beforehand, Kelly Slater made up his mind to journey to an all-inclusive resort called "Rhythmia" to experiment with Ayahuasca, an indigenous Amazonian psychedelic brew. Rhythmia is not just a vacation retreat, but consists of a demanding schedule of introspective and wellbeing practices, as well as indigenous rituals that focus on the ingestion of and healing with ayahuasca.

Shortly after his weeklong stay, Slater was featured in a video on the company's website describing his experience as "the most profound… of my life" (Rythmia Life Advancement Center, 2017). Later on, during an interview with mentor and professional surfer Tom Carroll—who suffered a long bout of addiction—Slater elaborated on the insight he derived from his psychedelic journey. He reflected on the wisdom he gained during his experience with the Amazonian brew: that what it takes to become the reigning world champion was that he had to sacrifice other parts of his life—family, friends, emotions, passions. The tenacious, uncontested competitive drive that produced in him a "heart of stone," he told Carroll, ultimately impacted negatively upon others close to him—though he did not specify who or how (Surfline, 2024).

For those familiar with Slater's long, prosperous career—whether as the most decorated surfer in history, a Hollywood figure, or as an entrepreneur and philanthropist—his admission of using drugs is quite surprising. This is because it is well known that Slater has largely steered clear of mind- and body-altering substances and has been public about his staunch admonishment of them. That philosophy is exemplified, for example, in a video that Slater once made for Natural High, a non-profit group that focuses on drug abuse prevention, where he admonished the use of all drugs and told a story about how he once threatened a guy with a bat who offered him LSD at a party (Natural High, 2019). Yet, people change, and the discourse around psychedelics today is largely positive, introspective, and promises a positive and transformative experience. Not least due to his positive experience, Slater has subsequently become a board member at Rhythmia.

Jamie Mitchell, too, has participated in the ayahuasca healing rituals at Rhythmia. Having had a few experiences with psychedelics prior to his trip to the Costa Rican resort, Mitchell describes partaking in around 30 ayahuasca ceremonies between Rhythmia and the jungles of Peru. In a 2022 podcast interview on *The Kyle Thiermann Show* (Thiermann, 2022a), Mitchell explained his use of psychedelics as an important tool for self-knowledge and a source of information, describing how it allowed him to make better decisions in his role as a father and husband. "You learn a lot about yourself […] I went pretty deep up there." In the discussion, Mitchell illustrates how transformative psychedelics have been for both he and his wife—that they've provided deep personal insights, including reflections on childhood, ego, and relationships, finally leading to a realization about the need to operate less selfishly and start prioritizing his wife's aspirations over his own.

While Mitchell summarizes his psychedelic experience as "awesome," he also highlights how it can be "brutal" at times and "should not be taken lightly." Both he and Slater articulate just how difficult it can be to narrate a clear description of any one experience—particularly for those who have not had a journey themselves. Yet for both Jamie as well as his wife, they believe the use of "plant medicine" as a tool of introspection, self-growth, and healing is worthwhile. "I just wanna be a better dad, a better friend, a better husband, a better person," Mitchell says. "That's what that is."

Resurfacing: Koa Smith's "Last Ditch Effort" To Heal

Another of the many high-profile surfers who have recently come out of the psychedelic closet is Koa Smith. Yet unlike Slater, whose ayahuasca use came after a long period of contemplation, for Smith, turning to psychedelics appeared to be the last available option before time ran out.

A few years back while surfing off the Indonesian island, Nias, Smith wiped out, brushed over the reef, hit his head, and was knocked unconscious. A traumatic brain injury (TBI) brought severe symptoms, the worst of which was a deep depression that led him to contemplate suicide. The documentary *Resurfacing* recounts his experience of hopelessness after being given no timeline for, nor promise of, recovery (MUDWTR, 2022a). "I could just end my life (shrugs)," he states in the documentary, recollecting his thought process at the darkest moments. "Na, I don't want to do that—don't do that. That's a dumb thought. But I *could* do it, that could get me out of this [situation]—*pretty quickly, actually*." After denying the antidepressant medication prescribed to him by a doctor, Smith figured he would rather try a natural remedy instead: taking, as he put it, "a large dose of mushrooms … [to] see if I could heal my head injury myself." It was a "last ditch effort [to survive]."

In the documentary, the audience follows Smith and his brother to a serene location, where he describes ingesting a large dose of psilocybin mushrooms. The trip, as he recalls it, was profoundly introspective, and he received a variety of important insights about his health, his TBI, and his early, rushed, and failed attempts to heal. He learned a powerful lesson on that day, one that seemed so obvious after the fact: he needed to listen to his body, as his hasty attempts to recovery quickly were actually worsening his symptoms and deepening his despair. "Let your brain be your guide to your life," he journaled after his psychedelic journey. "Let it lead you to your path to *healing*. Listen, check in with it, slow build, and become the superhuman you are." In this altered state, he gained a new understanding of his injury and was guided toward a more authentic and sustainable healing path.

Resurfacing brilliantly tells the story of Smith's (newfound) process of recovery. When the film opened at a release party in Santa Monica, a live viewing was presented, followed by a panel discussion that featured Smith, a clinical psychedelic researcher from the Multidisciplinary Association for Psychedelic Studies (MAPS) named Dorna Pourang, and professional big-wave surfer and podcaster Kyle Thiermann (MUDWTR, 2022b). A long conversation ensues with Smith answering questions, Pourang providing an overview of the therapeutic mechanisms offered by psychedelics, and Thiermann moderating and offering his own personal insight. At the outset of the panel, Thiermann describes the current era as "the second psychedelic revolution," predicting it will be "one of the greatest positive changes we see in society in our lifetime." As he goes on, he frames the psychedelic renaissance through the lens of medicalization. In 1970, Thiermann explains to the audience, a "sweeping ban … really halted a lot of research." But the current "mental health epidemic," and a broader societal shift towards "destigmatizing" psychedelics, has led to "one of the few bipartisan issues" of our time, he proclaims.

"My whole goal with this film," Smith stated later on during an interview on Thiermann's podcast, "is letting people know how profound my experience was" (Thiermann, 2022c).

> Through the psychedelic experience, it cured my deep depression and my brain injury in one setting. I think a lot of people are gonna be relating to what I was going through. This life is so special and so unique, that it's so worth living. And I hope this (documentary) just inspires people deep down to know that they can heal.

Smith, like Slater and Mitchell, turned to psychedelics for their therapeutic potential, though he did so in a different context. Struggling with depression and the challenges of a TBI, he used psychedelics intentionally as a healing tool. This approach provided him with deep introspection and insight, allowing him to integrate the lessons from his experiences into daily life and support his overall healing process.

Jon Rose: Humanitarian Work and Post-traumatic Stress

For Jon Rose, former professional surfer and founder of the non-profit group Waves for Water, doing humanitarian work around the world comes at a cost. Spending much of his time the last several years traveling to developing countries in search of bringing clearn water to deprived communities, Rose has encountered a number of situations and circumstances that have left a lasting mark on his psyche. For example, during an interview in 2022 (Thiermann, 2022b), Rose recounted in detail just one of several harrowing incidents he's experienced in his line of work. While in the favelas in Brazil, after wandering into a nearby yard to recover a momentarily displaced soccer ball, he encountered a man attempting to rape a child. Acting heroically, Rose intervened physically, preventing the act from taking place. Yet, the weight of this experience (not to mention several other traumatic encounters he has yet to discuss in public) continues to deeply affect him—the lingering memories and the ineffable nature of trauma reemerging unexpectedly, causing significant emotional distress (Rose, 2024a).

Though Rose has sought treatment through talk therapy, it has proven mostly ineffective, leaving him with a desire to attempt other, non-traditional means. He is also well aware of the stigma surrounding men discussing their emotions, a topic he revealed struggling with during the interview. Having experimented with psychedelics recreationally on numerous occasions, yet over time learning more about the therapeutic value they can offer, Rose now feels ready to take a more intentional and structured approach through psychedelic-assisted therapy. In the interview, he distinguishes between recreational use—characterized by spontaneity and less emphasis on set, setting, and therapeutic intent than the medicalized approach—and the potential of psychedelics when used responsibly and for the purposes of healing, describing how he plans to utilize the latter soon. The promise of PAT, Rose says, is about "not just releasing, but understanding some of the things that I've experienced, and ultimately, processing them."

At the time of this writing, it is unclear if Rose has used psychedelics as an adjunct to therapy. Yet his discourse nevertheless perpetuates the knowledge that such substances are worthwhile in surf and broader culture, that there is a difference between recreational and therapeutic experiences, and that it is acceptable for men to dig deep into their psychological and emotional lives as a means of revealing and healing trauma.

Kohl Christensen and the Ain't That Swell Podcast: Soul Ain't Over Just Yet

With all the discourse about the therapeutic value of psychedelics in mainstream culture, it would be a mistake—and, let's be honest, *a bit of a shame*—to say that today's world of surfing has completely lost its "soul." The good thing is that there are still some prominent voices in the culture who are keeping the past funky, free-spirited era alive, passing down the stories, practices, and philosophies from the soul tradition—and who take psychedelics in a spiritual, countercultural, and even in deliberately "irresponsible" manner (at least by the more conservative advocates of the medicalized approach). One of them is big-wave charger Kohl Christensen.

In 2023, Christensen was interviewed on the popular Australian surf podcast *Ain't That Swell*—a soul-inspired show in and of itself, whose main host, Jed Smith ("Smivvy"), announces himself at the outset of each episode as the "Punch Drunk Pikey," the "Sultan of Psilocybin," and the "Maestro of Microdosing." During their discussion, Christensen recounted his varied experiences with psychedelics and other drugs, from intentional ayahuasca retreats and structured rituals to more spontaneous encounters with LSD, mushrooms, and the Belladonna flower, datura—the latter he describes as inducing a 24-hour experience that was "on the extreme side of psychedelics, where you actually detach." Throughout the podcast, it becomes clear that both Smivvy and Christensen are well-versed in the soul tradition and Sixties folklore, delving into, amongst several other countercultural topics, figures like Timothy Leary and his colleague Ram Dass (who later turned away from psychedelics to meditation and other non-drug spiritual practices). When Christensen mentions Leary and his written works, Smith excitedly and eagerly interjects, "*The Psychedelic Experience*, the book they've written! That's a great book! That's basically the fuckin' duck's nuts in psychedelic protocol" (Ain't That Swell, 2023).

Though both are steeped in the old-school soul tradition and well-versed in 1960s psychedelic history, Christensen and Smivvy take a more nuanced stance and can appreciate the medicalized psychedelic culture currently permeating surfing and the broader society—a phenomenon that clearly stretches outside the mainland to both Hawaii as well as to Australia. Throughout the discussion it becomes clear that both parties' hold a unique understanding of the distinct drug-taking traditions. At one point in the interview, Christensen acknowledges the importance of the more contemporary approach to psychedelics, admitting he's undergone a "learning curve," moving beyond the fun and chaos of classic psychedelic use to recognizing the value of intentional, therapeutic experiences. "It's not necessarily about having a great time," he says, referring to the recreational style of use. "That's a byproduct. The real focus is how psychedelics allow you to change your perspective." Recognizing the broader cultural shift from soul to the medicalized approach, he describes the importance of set, setting, and working with sitters, elders, or guides. His advice to others? "Talk to friends, ask about their experiences, and read some books."

Thus, as Christensen and shows like *Ain't That Swell* demonstrate, the soul tradition is far from over—and many of us can be thankful for that. However, what is clear is that even the soul tradition has become influenced by a more professionalized, responsible, medicalized approach to the use of psychedelics—the demarcation of which can be clearly understood in podcasts such as these where the pros and cons of both traditions are discussed and contrasted at length.

The (Re)Shaping of Psychedelic Surf Discourse

As these brief yet impactful examples show, professional surfers are stepping out of the psychedelic closet, openly sharing their experiences and pushing psychedelics further into various segments of mainstream culture. Unlike the spiritual and countercultural associations of yesteryear—and their political implications—today we are witnessing a shift in the discourse surrounding psychedelics. For the most part, they are now framed as therapeutics and "plant medicines," recognized for their potential to heal a broad spectrum of health issues, ranging from physical and psychological to relational (Rose, 2025). In the current era, much of the spiritual components of the psychedelic experience have been eschewed to adhere to conventional tastes and belief systems (though, today, even terms like "mystical experience" have scientific backing - see Griffiths et al. 2006).

Professional surfers, due to their status and large audiences, play an active role in shaping the psychedelic discourse. Their stories not only admit to psychedelic use but also advocate for its philosophy, influencing cultural norms and values. These public discussions do more than inform—they establish authority, reinforce behavioral standards, and frame drug use, health, and well-being in therapeutic terms. By positioning psychedelics as tools for mental health and personal growth, surfers are helping to define the future of psychedelic use, both within and beyond surf culture. As the psychedelic movement continues to expand into new areas and corporate ventures, and with psychedelic retreats increasing each year, history may ultimately reveal that we are only witnessing the early stages of the psychedelic renaissance. While it is possible that the soul tradition—with its subversive, divine, and radical implications—comes back into relief, it is quite likely that for the foreseeable future, the culture of surfing will follow the many pathbreakers of recent who have experimented with PAT and shared their stories.

Taking this current discourse into consideration—*the shift from soul to science*—let us take a quick glance at some of the ways this cultural turn might impact the future of surfing.

From Soul to Science: The Impact of Psychedelics on the Future of Surfing

Psychedelics—once tied to counterculture, hippies, and revolutionary ideals—have entered the mainstream. Today, the focus is no longer on turning on, tuning in, and dropping out, but on healing and self-improvement. Though still largely prohibited by law in most Western countries, this reality is rapidly shifting with decriminalization and medicalization. Given how deeply psychedelic culture has influenced both past and present surf eras, how might this ongoing shift toward medicalization shape the future of surfing?

The Impact of Psychedelic Medicalization

On the one hand, the practices surrounding psychedelics are changing. As we've seen, when psychedelics are advocated for therapeutic use, the goal is to explore the inner landscape and enhance awareness of both conscious and unconscious patterns. By directing the drug, and concomitantly, broader surf culture, toward this objective—which encourages self-reflection and -analysis and the examination of behavioral, psychological, and relational patterns—important implications will naturally arise for health, well-being, and personal change. In other words, just as the soul surfing tradition shaped surfers' personal practices and philosophies (toward mystical and anti-establishment ends), today's shift will introduce new practices and beliefs that will permeate the culture, shaping perspectives on health and well-being (as explored in the sections below).

Yet what about the experiential consequences engendered by such practices? In the same manner that the medicalized psychedelic framework promotes a more mindful effort to examine the self, it is important to consider the role that the *phenomenological,* or experiential content of psychedelics will have on the culture. Many clinical studies have explored the direct impact of phenomenological qualities unique to psychedelic experiences, particularly when the experience is contemplative or self-directed. For example, studies have found that participants of PAT can achieve an increased sense of ego dissolution (a brief state of diminished self-reference – see Kałużna et al., 2022), universal connectedness (a sense of infinite unity with others or the environment – see Carhart-Harris et al., 2018), increased nature relatedness (an enhanced sense of connection to the natural environment – see Kettner et al., 2019), and an improved closeness with others in group-based psychedelic contexts (see Rose, 2025), amongst other experiences. Each of these, notably, is correlated with positive mental health outcomes (Kałużna et al., 2022).

Will surfing go through (another) egoless era, espousing connection, increasing nature relatedness, and championing discussions about mental (or spiritual) health and wellbeing? Will the tide shift towards something more adjacent to a Soul Surfing 2.0? (If this all feels too idealistic—that psychedelics only cultivate peace-loving hippies—let me remind you that, as history demonstrates, Nazis use psychedelics too (see Pace & Devenot, 2021)). Hence, with the increasing medicalization of psychedelic practices and philosophies, surf culture will continue to find itself infused with attention to holistic and spiritual forms of betterment and personal healing practices (the "checking in" with the self that Koa Smith promoted, or the self-knowledge espoused by Slater and Mitchell).

The Destigmatization of Mental and Emotional Health

As a byproduct of the medicalized psychedelic renaissance, we are seeing public discussion of topics that have historically been stigmatized, such as mental and emotional health (particularly amongst men), take on a new primacy in the culture. This is important for a number of reasons. While historically, medicine has treated the mind as separate from the body, we now know much more about the interconnectedness of the mind and body, and its situatedness in a broader sociocultural and political environment. From discussions of stress, anxiety, and depression to trauma, we recognize more than ever that our well-being is deeply relational—that is, connected to our emotions, our bodies, and others. For several important reasons, the re-emergence of psychedelics in the twenty-first century is helping to validate these historically marginalized ideas, paving the way for a truly biopsychosocial science of health and well-being.

An important aspect of this broader transition to a more holistic style of health is the normalization of talking about mental and emotional health. Men have been shunned for speaking about such topics because of the stigma surrounding them. While often imposed from the outside, stigma can also originate internally, resulting in feelings of shame or inadequacy. Also problematic is that the fear of stigma can result in men—who are taught to always portray strength, self-reliance, and capability—hiding symptoms or not seeking mental healthcare, thus potentially increasing isolation, suffering, and self-harm (McKenzie et al., 2022). As the exploration of one's psychic life through psychedelic drugs becomes more common, conversations about distress, anxiety, depression, and trauma do too, giving men the opportunity to talk about their feelings, worries, shame, and sadness without fear of retribution. Ultimately, what this means is a broader turn toward mental health awareness. And as we've seen, surfers, when contributing to broader cultural discourse, will impact the lives of many with their narratives of healing, self-awareness, and therapeutics.

From Individualism to Collectivism

As we have seen, each individual mentioned above (aside from Smith) has partaken in a psychedelic experience through collective ceremonies in retreat settings. Since the 1980s, there has been a growth in psychedelic retreats (often described as "shamanic tourism") attempting to meet a growing demand for facilitated drug experiences. While clinical trials focus on individual-level care with tightly controlled variables, most people won't have access to the limited spots available in these studies. As a result, the future of psychedelics is likely to unfold primarily within group-based retreats. This shift is particularly significant not only for the therapeutic potential it offers, but also for the economic benefits tied to the growth of group-based retreats (Marseille et al., 2023).

In Western societies, individualism prevails, even though there is a deep, shared need for community. These cultures, especially as shaped by neoliberal political and economic frameworks, often frame mental health and wellbeing as an individual responsibility. The use of psychedelics in community will hopefully contribute to offsetting this despairing climate of isolation and loneliness. A number of important connective and healing dynamics arise in such communal contexts—what I, following the classical sociologist Emile Durkheim, have called "collective effervescence" (Rose, 2025).

Connectedness, in this sense—not just in the phenomenological way previously discussed—is crucial for health, wellbeing, and building supportive communities. The soul surfers were keenly aware of the power of community, particularly under the influence of the Brotherhood, who were as group-oriented as it gets. They "turned on" thousands to the communal ethos, fostering a collective spirit that was far more pronounced than what we see today. In contrast, the current landscape of surfing, influenced heavily by the monetization of social media, allows surfers to make money through individual advertising deals, often at the cost of communal bonds—both with benefits and drawbacks. It's worth considering whether the future of surf culture might place more emphasis on the collective, offering a counterpoint to the increasingly isolated existence many of us are experiencing today. This would be my hope, at least.

From Abuse to Enlightened Forms of Drug Use

With the rise of PAT, along with the growing destigmatization of mental health and a shift toward collectivism, a natural next step is a more open and honest conversation about the benefits and drawbacks of drug use. Many unfamiliar with the sport and culture of surfing may look at the question of drug use with skepticism—assuming that surfers lead inherently healthy, substance-free lifestyles. The reality, however, is far more complex. From Hawaii to California, Australia, and otherwise, many surfers have been damaged by the culture's history of drug problem. And in some ways, the culture of surfing itself—which at times, for example, allows beer companies to sponsor competitions—has even advocated a recreational drug use and party-like atmosphere. From Andy and Bruce Irons to Tom Carroll, Darryl "Flea" Virostko, and others, there have been some notorious cases of substance abuse. And while many have been fortunate to make it out alive, others like Andy weren't so lucky.

Surfing naturally promotes in some of its adherents a lifestyle that contributes to an increased likelihood of addiction (Grodesky, 2014). Surfers are naturally drawn to exceptional, mind-bending experiences, influenced by nature's raw power, exploring new territories, and navigating the unpredictable sea. This adventurous lifestyle, often coupled with the adrenaline rush of professional surfing, creates a constant pursuit of dopamine highs. However, amidst these thrills, other challenges—emotional, psychological, and social—can complicate their journey.

For Carroll, for example, the high-intensity, status-driven lifestyle of a professional surfer in the limelight became difficult to handle. Not to mention the fact that he lost his mother at the age of seven, and later on his sister. Even at the height of his professional career, feelings of inadequacy overwhelmed him. As he claimed during his *60 Minutes* interview, his methamphetamine addiction led him to feel "so hollow emotionally. I wasn't connecting to anyone anymore, other than the dealer and the drug." As well, Bruce Irons, after the loss of his brother Andy, turned even further towards substance abuse. Just last year in 2024, however, Bruce, too, recently opened up as having used psychedelics for help, experimenting with the African ibogaine plant at a retreat in Mexico to treat his substance abuse and mental health issues. And some like Flea are using their own experiences to help others navigate addiction. He has opened up a rehabilitation center called "The Fleahab House" in Santa Cruz, California.

It is often said that culture is a reaction to that which came before it, and for many who have been negatively impacted by surfing's drug problem, they wish the next generation would make better choices. As today's surfers openly discuss their use of psychedelics for therapeutic purposes—sharing stories of past mistakes, excessive partying, and the consequences of those choices—they are offering valuable lessons for both their peers and future generations. This openness will help shape the way contemporary surfers and those to come approach their own journeys—both in surfing and in life.

Currently, there are many reasons even beyond the rise of psychedelic therapy that suggest a change in how surfers approach drug and alcohol use compared with their predecessors in the sport. There is, for example, a growing cultural shift away from alcohol, driven by scientific discourse revealing its more harmful effects, including recent moves by the U.S. Surgeon General to classify alcohol as a carcinogen. Simultaneously, there's an increasing focus on the negative aspects of problematic drug use, with figures like Andrew Huberman, for example, using social media to highlight the detrimental impacts of substances like alcohol. Recent studies, for instance, challenge previous cultural beliefs, showing that even moderate alcohol consumption—one to two drinks daily—can harm brain function (Daviet et al., 2022). In contrast, some academics openly advocate for responsible drug use, with Carl Hart's *Drug Use for Grown-Ups* (Hart, 2021) being a leading resource. This middle ground—between total sobriety and substance abuse—may appeal to some, as PAT may foster the introspection needed to make more informed decisions about drug use.

Overall, it's important to recognize that conversations surrounding the therapeutic benefits of psychedelics—distinct from the countercultural and oftentimes chaotic uses of the past—are becoming deeply embedded within surf culture, as seen in the aforementioned conversations about PAT. This shift, fueled by the growing focus on mental and emotional health, is likely to gain momentum in the future.

Conclusion

Psychedelic drugs have long been intertwined with the history of surfing, playing a significant role in shaping its countercultural ethos. Today, however, a more medicalized approach to drug use and experiences has overtaken the spiritual, recreational, and at times irresponsible consumption that characterized the soul surfing era. This shift is evidenced by the growing number of surfers who embrace a framework for psychedelic use that emphasizes health, well-being, and scientific understanding. Looking ahead, the future of surf culture seems poised to prioritize holistic health and well-being in a more biopsychosocial context than ever before. As psychedelics continue to make their way into surf culture, we can expect an increasing focus on the destigmatization of mental and emotional health, especially for men, a renewed emphasis on community and collectivity (both within psychedelic ceremonies and beyond), and a broader recognition of safer, more intentional drug use practices compared to the past. This second wave of psychedelics in surfing culture offers a promising future where health, connection, and personal growth take center stage—a transition from soul to science.

References

Aint That Swell (Director). (2023, September 22). *Core Lords: Kohl Christensen* [Video recording]. https://www.youtube.com/watch?v=8RR2lfyJC5g

Bogenschutz, M. P., Podrebarac, S. K., Duane, J. H., Amegadzie, S. S., Malone, T. C., Owens, L. T., Ross, S., & Mennenga, S. E. (2018). Clinical Interpretations of Patient Experience in a Trial of Psilocybin-Assisted Psychotherapy for Alcohol Use Disorder. *Frontiers in Pharmacology, 9.* https://www.frontiersin.org/article/10.3389/fphar.2018.00100

Booth, D. (1994). Surfing'60s: A case study in the history of pleasure and discipline. *Australian Historical Studies, 26*(103), 262–279. https://doi.org/10.1080/10314619408595963

Booth, D. (1995). Ambiguities in Pleasure and Discipline: The Development of Competitive Surfing. *Journal of Sport History, 22*(3), 189–206.

Booth, D. (1996). Surfing Films and Videos: Adolescent Fun, Alternative Lifestyle. *Adventure Industry. Journal of Sport History, 23*(3), 313–327.

Carhart-Harris, R. L., Bolstridge, M., Rucker, J., Day, C. M. J., Erritzoe, D., Kaelen, M., Bloomfield, M., Rickard, J. A., Forbes, B., Feilding, A., Taylor, D., Pilling, S., Curran, V. H., & Nutt, D. J. (2016). Psilocybin with psychological support for treatment-resistant depression: An open-label feasibility study. *The Lancet Psychiatry, 3*(7), 619–627. https://doi.org/10.1016/S2215-0366(16)30065-7

Carhart-Harris, R. L., Erritzoe, D., Haijen, E., Kaelen, M., & Watts, R. (2018). Psychedelics and Connectedness. *Psychopharmacology, 235*(2):547–550. https://doi.org/10.1007/s00213-017-4701-y

Daviet, R., Aydogan, G., Jagannathan, K., Spilka, N., Koellinger, P. D., Kranzler, H. R., Nave, G., & Wetherill R. R. (2022). Associations between alcohol consumption and gray and white matter volumes in the UK biobank. *Nature Communications, 13*(1):1175. https://doi.org/10.1038/s41467-022-28735-5

Garcia-Romeu, A., Griffiths, R., & R., & W. Johnson, M. (2014). Psilocybin-Occasioned Mystical Experiences in the Treatment of Tobacco Addiction. *Current Drug Abuse Reviews, 7*(3), 157–164.

Giffort, D. (2020). *Acid Revival: The Psychedelic Renaissance and the Quest for Medical Legitimacy*. University of Minnesota Press. https://doi.org/10.5749/j.ctv13qfvv7

Griffiths, R. R., Johnson, M. W., Carducci, M. A., Umbricht, A., Richards, W. A., Richards, B. D., Cosimano, M. P., & Klinedinst, M. A. (2016). Psilocybin produces substantial and sustained decreases in depression and anxiety in patients with life-threatening cancer: A randomized double-blind trial. *Journal of Psychopharmacology, 30*(12), 1181–1197. https://doi.org/10.1177/0269881116675513

Griffiths, R. R., Richards, W. A., McCann, U., & Jesse, R. (2006). Psilocybin can occasion mystical-type experiences having substantial and sustained personal meaning and spiritual significance. *Psychopharmacology, 187*(3), 268–283. https://doi.org/10.1007/s00213-006-0457-5

Grodesky, N. (2014, February 4). Under the Influence of Surf Culture and Understanding Addiction. https://www.theinertia.com/surf/under-the-influence-of-surf-culture-and-understanding-addiction/

Hart, C. L. (2021). *Drug Use for Grown-Ups*. Chasing Liberty in the Land of Fear. Penguin Publishing Group.

Hartogsohn, I. (2022). Modalities of the psychedelic experience: Microclimates of set and setting in hallucinogen research and culture. *Transcultural Psychiatry, 59*(5), 579–591. https://doi.org/10.1177/13634615221100385

Johnson, M. W., Garcia-Romeu, A., Cosimano, M. P., & Griffiths, R. R. (2014). Pilot study of the 5-HT2AR agonist psilocybin in the treatment of tobacco addiction. *Journal of Psychopharmacology, 28*(11), 983–992. https://doi.org/10.1177/0269881114548296

Johnson, M. W., Garcia-Romeu, A., & Griffiths, R. R. (2017). Long-term follow-up of psilocybin-facilitated smoking cessation. *The American Journal of Drug and Alcohol Abuse, 43*(1), 55–60. https://doi.org/10.3109/00952990.2016.1170135

Kałużna, A., Schlosser, M., Craste, E. G., Stroud, J., & Cooke, J. (2022). Being no one, being One: The role of ego-dissolution and connectedness in the therapeutic effects of psychedelic experience. *Journal of Psychedelic Studies, 6*(2), 111–136. https://doi.org/10.1556/2054.2022.00199

Kettner, H., Gandy, S., Haijen, E. C. H. M., & Carhart-Harris, R. L. (2019). From Egoism to Ecoism: Psychedelics Increase Nature Relatedness in a State-Mediated and Context-Dependent Manner. *International Journal of Environmental Research and Public Health, 16*(24). https://doi.org/10.3390/ijerph16245147

Lawler, K. (2011). *The American surfer: Radical culture and capitalism* [Electronic resource]. Routledge. http://myaccess.library.utoronto.ca/login?https://www.taylorfrancis.com/books/9780203838754

Leary, T., Metzner, R., & Alpert, R. (1969). *The Psychedelic Experience: A Manual Based on the Tibetan Book of the Dead*. Psychedelic Monograph. University Books.

Marseille, E., Stauffer, C. S., Agrawal, M., Thambi, P., Roddy, K., Mithoefer, M., Bertozzi, S. M., & Kahn, J. G. (2023). Group psychedelic therapy: Empirical estimates of cost-savings and improved access. *Frontiers in Psychiatry, 14*. https://doi.org/10.3389/fpsyt.2023.1293243

McKenzie, S. K., Oliffe, J. L., Black, A., & Collings, S. (2022). Men's Experiences of Mental Illness Stigma Across the Lifespan: A Scoping Review. *American Journal of Men's Health, 16*(1), 15579883221074789. https://doi.org/10.1177/15579883221074789

Moreno, F. A., Wiegand, C. B., Keolani Taitano, E., & Delgado, P. L. (2006). Safety, Tolerability, and Efficacy of Psilocybin in 9 Patients With Obsessive-Compulsive Disorder. *The Journal of Clinical Psychiatry, 67*(11), 1735–1740. https://doi.org/10.4088/JCP.v67n1110

MUDWTR (Director). (2022a, March 17). *Resurfacing: Koa Smith* [Video recording]. https://www.youtube.com/watch?v=zKPF7r-pMsI

MUDWTR (Director). (2022b, March 17). *"Resurfacing: Koa Smith" Q&A Panel* [Video recording]. https://www.youtube.com/watch?v=qjySHkd0w8U

Natural High (Director). (2019, May 14). *Kelly Slater, "Greatest Surfer of All Time", talks living a drug-free life* [Video recording]. https://www.youtube.com/watch?v=KLHffQPGTZM

Neitzke-Spruill, L. (2023). *Psychedelic biomedicalization mainstreaming a scientific revolution* [University of Delaware]. https://udspace.udel.edu/server/api/core/bitstreams/0ea72151-d46d-457b-852a-ce2050de0282/content

Pace, B. A., & Devenot, N. (2021). Right-wing psychedelia: case studies in cultural plasticity and political pluripotency. *Frontiers in Psychology, 12.* https://www.frontiersin.org/articles/10.3389/fpsyg.2021.733185

Rose, J. R. (2024a). Memory, trauma, and self: Remembering and recovering from sexual abuse in psychedelic-assisted therapy. *Journal of Psychedelic Studies, 8*(3), 336–348. https://doi.org/10.1556/2054.2024.00363

Rose, J. R. (2024b). Turn On, Tune In, and Paddle Out: A Cultural Theory of Psychedelic Amplification and Coproduction in the Soul Surfing Era. In M. J. Roberts, K. Lawler, & D. P. Cline (Eds.), *Roll and Flow: The Cultural Politics of Skateboarding and Surfing.* San Diego State University Press.

Rose, J. R. (2025). *From Isolation to Social (and Psychedelic) Integration: A Sociology of Collective Effervescence and Therapeutic Community in Psychedelic-Assisted Therapy.* Sociology of Health & Illness.

Roseman, L., Nutt, D. J., & Carhart-Harris, R. L. (2018). Quality of Acute Psychedelic Experience Predicts Therapeutic Efficacy of Psilocybin for Treatment-Resistant Depression. *Frontiers in Pharmacology, 8.* https://doi.org/10.3389/fphar.2017.00974

Ross, S., Bossis, A., Guss, J., Agin-Liebes, G., Malone, T., Cohen, B., Mennenga, S. E., Belser, A., Kalliontzi, K., Babb, J., Su, Z., Corby, P., & Schmidt, B. L. (2016). Rapid and sustained symptom reduction following psilocybin treatment for anxiety and depression in patients with life-threatening cancer: A randomized controlled trial. *Journal of Psychopharmacology, 30*(12), 1165–1180. https://doi.org/10.1177/0269881116675512

Rythmia Life Advancement Center (Director). (2017, November 30). *Kelly Slater Ayahuasca Experience: "Otherworldly"* [Video recording]. https://www.youtube.com/watch?v=QqJzyL0wWPg

Surfline (Director). (2024, April 8). *Kelly Slater: Retirement, fatherhood, and a champion's changing heart.* [Video recording]. https://www.youtube.com/watch?v=KwjQftM8yj4

Thiermann, K. (2022a, May 18). #276 Ayahuasca & Beat Downs—Jamie Mitchell [Substack newsletter]. *Kyle Thiermann.* https://thiermann.substack.com/p/276-ayahuasca-and-beat-downs-jamie

Thiermann, K. (2022b, May 18). #290 Psychedelics, PTSD, & the Haiti Earthquake—Jon Rose [Substack newsletter]. *Kyle Thiermann.* https://thiermann.substack.com/p/290-psychedelics-ptsd-and-the-haiti

Thiermann, K. (2022c, May 18). #303 Resurfacing with Koa Smith and Chris Keener [Substack newsletter]. *Kyle Thiermann.* https://thiermann.substack.com/p/303-resurfacing-with-koa-smith-and

Watts, R., Day, C., Krzanowski, J., Nutt, D., & Carhart-Harris, R. (2017). Patients' accounts of increased "connectedness" and "acceptance" after psilocybin for treatment-resistant depression. *The Journal of Humanistic Psychology, 57*(5), 520–564. https://doi.org/10.1177/0022167817709585

Chapter 11
Data Bitches, Citizen Science, and Sea-Level Rise

Briana M. Iatarola

Most surfers recognize sea-level rise (SLR) poses a major threat to vulnerable surf breaks. One of the most famous and heavily visited surf spots in Southern California is Trestles, which is also vulnerable to SLR. The creeks of the San Mateo Point watershed empty into the Pacific Ocean at the intersection of San Onofre State Beach and Marine Corps Base Camp Pendleton. Trestles' waves crash against 20 acres of beach, and "San O" contains 174 acres of public park space between Los Angeles and San Diego and within proximity to 65,000 residents. Trestles' waves are also considered among the best on this planet. A rich coupling of economic, sociocultural and ecological attributes also makes Trestles a dynamic site to conduct research, particularly because SLR has already affected San Onofre State Beach. A growing number of community science initiatives aim to provide digitally democratic ways for surfers and non-scientists to document the impacts of SLR at vulnerable surf breaks.

In December 2015, program organizers issued a call for "citizen scientists" concerned about SLR and the effects of El Niño along California's coast to join the University of Southern California's National Sea Grant College Program (USC Sea Grant). SLR was the pressing issue that USC Sea Grant addressed through the "Urban Tides Community Science Initiative" (Urban Tides). Researchers geo-referenced a list of beaches in need of data about "current tidal lines, beach erosion, and coastal flooding," and San Onofre State Beach was one site of interest.[1] Given my concerns about the impact of SLR on California's beaches, I answered the call in 2016 as a doctoral student in the Department of Communication at the University of California,

[1] See: Urban Tides Community Science Initiative advertising flier.

B. M. Iatarola
University of California San Diego, San Diego, USA

B. M. Iatarola (✉)
San Diego State University, San Diego, USA
e-mail: biatarol@ucsd.edu

L. Mach (ed.), *The Future of Surfing in the Anthropocene*,
https://doi.org/10.1007/978-3-032-09377-6_11

San Diego. One of my research objectives involved exploring the tenuous outcome of the Save Trestles campaign to stop the controversial 241 Toll Road extension. I was keen to see how incorporating the Urban Tides initiative's research methodology would alter, amplify, or change my own understanding of SLR data collection.

In the case of the Urban Tides citizen science project, three technological tools were central to SLR data collection efforts along Southern California's coast in 2016. They included a user's smartphone, an SLR application (a.k.a. app), and a cloud-based storage system designed by the now-defunct start-up Liquid. Caleb LeNoir and Jacob Shidler founded the data-management company compatible with mobile devices in 2013; they collaborated to build a database-management system that was "basically an electronic, online version of the scientific method" (Smith, 2015).[2] Shidler described data as "siloed in the sciences because of this paper-based system. It's not fluid and transferable across disciplines, across researchers, across departments, even" (via Smith, 2015). Despite LeNoir and Shidler's goal to democratize the process of data collection, by June 1, 2018, the Liquid platform shut down. In this chapter, I provide a critical reflection on what it means to be a citizen scientist using a smartphone and app as participatory tools for data collection. Reflecting on my yearlong involvement in Urban Tides throughout 2016, I identify methodological challenges that arose when I relied on my iPhone 6s to collect SLR data with Liquid's app and sync it to the start-up's server. In documenting SLR conditions at Trestles in San Onofre State Beach during 2016 El Niño, syncing issues ultimately rendered my data unusable for scientific analysis. As such, researchers and academic institutions that collaborate with private start-ups that provide a cloud-based storage system for valuable coastal data are engaging in a fraught process. Data extinction is a real possibility if the technological tools are compromised, or the start-up does not survive.

Cultivating public-private research partnerships is vital for advancing citizen science projects, and volunteers often play a key role in the data-collection process. Coastal research that has turned to surfers as data-collecting participants is evolving in step with the deployment of emerging technologies beyond smartphones and apps. Bresnahan et al.'s ongoing Smartfin project, for instance, demonstrates significant advancements in a surfing device engineered and distributed to encourage public participation in environmental research. In Smartfin's design, an "oceanographic sensor-equipped surfboard fin" serves as an integral tool for doing citizen science to produce "near-real-time data communications."[3] Collaborative endeavors with various stakeholders, including nonprofit and campus organizations, as well as university departments and affiliated research-oriented institutes, have "[...] proven to be a powerful way to combine the collection of coastal physical data with a citizen science project geared toward public participation in data collection, public outreach, and informal education" (Bresnahan et al., 2022, 8). The Smartfin project's "free and

[2] The start-up was based in Queensgate, Ohio.

[3] Bresnahan et al., explain (2022): "The Smartfin device is designed to collect temperature, position, and motion data from natural waters while matching performance specifications of a standard surfboard fin" (2).

open data-driven web portal" also has helped facilitate practical pathways for student research (Bresnahan et al., 2022, 2).

My research opportunity and experience with Urban Tides in comparison provokes two reflective questions: How might traditional or alternative methods for documenting SLR ensure valuable community science data does not disappear? And what are potential best practices to consider for future community science projects that turn to surfers to collect SLR data? If such data is supposed to inform future coastal resilience and adaptation strategies, then it must be preserved and remain publicly accessible to communities even after the project concludes or the server and platform shut down. After my community science experience, I no longer saw myself as a citizen scientist; rather, I felt more like a "data bitch" (Liboiron, 2017). I argue that being a data bitch in citizen science projects for scientists and start-ups conducting SLR research on the Anthropocene Ocean can inform ocean and climate change literacy at personal and local levels.[4] The erasure of public access to the 2016 Urban Tides data sets and demise of the novel Liquid app, however, reveal broader socio-technological implications for future citizen science projects that ask surfers to submit data about vulnerable surf breaks with a smartphone, app, and similar technologies. For these reasons, I conclude this chapter by providing three possible best practices for how administers of similar projects might overcome basic socio-technological challenges.

Citizen science at a rudimentary level involves the participation of non-scientists in the process of collecting scientific data for large-scale spatial and temporal projects (Grace-McCaskey et al., 2019). There is a broad range of practice that falls under the term citizen science. Critical literature points to poles of this spectrum where citizen participants have the following: a greater or lesser substantive involvement in determining research questions and design; access to data and the potential to participate in analysis; and a say at the table in terms of how findings are used and disseminated.[5] Although this chapter incorporates these themes, it does not analyze whether Urban Tides is a case study of citizen science "done right." This would be difficult considering the 2016 initiative was a socially minded, educational, and ideological project grounded in democratizing the production of scientific knowledge. I found myself questioning the notion of citizen science as democratizing science within the context of this specific initiative and my engagement with it. I thus analyze Urban Tides through a more critical lens.

In a less critical framing, citizen scientists are more generally described as volunteers who collect data for scientists who are conducting scientific research to broaden their understanding of human-environment relationships. But citizen scientists may also be thought of as data bitches. I attribute this quip to feminist environmental

[4] I recognize the scholarly disputes around "citizen science" and "problematic associations with the word 'citizen'" (Bresnahan et al., 2022; Cooper et al., 2012). I use the terms community science and citizen science interchangeably here based on recruitment language for Urban Tides volunteers.

[5] Rajul E. Pandya deconstructs several models of citizen science, explaining that income, job responsibilities, and lack of access to nature and transportation are "root causes for the lack of minority participation in citizen science." He encourages scientists to mine the published materials of long-term projects to understand "the dearth" (2012, 314–315).

scientist Max Liboiron, who responded to my question about their views on citizen science in an April 2017 "Designing Feminist Technologies" workshop at UCSD. By this point, I had concluded all fieldwork and was reflecting on my methodological approach to data collection. Conceiving of citizen science as primarily data collection and citizen scientists as data bitches is a very limited and depoliticized understanding of the terms and their histories. All etymology and pejorative word roots aside, Liboiron's blunt synonym for citizen scientists resonated. Urban Tides was an activist intervention into the fate of coastal communities. I was a data bitch driven by the desire to explore new means of co-producing knowledge about Trestles and San Onofre State Beach, knowing full well that I was environmentally opposed to the 241 Toll Road extension.

Urban Tides' model promotes an efficient "transfer of science-based information" rooted in applied research, community-based knowledge, environmental literacy, and communication. Participation in the initiative was voluntary and provided a way for people to engage in the multi-level fight to protect the ecological vitality of beaches. Participating in Urban Tides became one of my methodologies by which to view how Trestles as a space is produced, understood, interpreted, and negotiated.[6] As I took pictures of high and low tidal lines at San Onofre State Beach between January 2016 and January 2017, I quickly realized that cross-fertilizing citizen science with the SLR knowledge I derived as a surfer would widen the scale of my analysis. While citizen science broadened my scale of analysis in certain respects, it also limited my observations. I found myself drawing on my surfer's knowledge and observations to provide cultural and ecological context for more unidimensional spatial models that informed how I gathered data, what data I gathered for Urban Tides, and how I transferred and communicated that data into knowledge.

What followed then was an experiment in this cross-fertilization of mixed methods that involved an "eco-ethnography" (Grace-McCaskey et al., 2019). This experimental method responds to Bonney and Dickinson's analysis of citizen science as a method and their call for "new language to talk across the disciplines, new models for understanding project design and project results, and new tools to bring people, nature, and computers together in meaningful ways" (2012, 26). To make sense of and navigate these spatial complexities as a data bitch, academic researcher and surfer, I conducted an eco-ethnography of Urban Tides, within which a data bitch functions as one more agent, and citizen science as another practice deployed in the production

[6] As an academic researcher, and surfer, I was gathering several types of data that appealed to very different audiences and wanted to become critically involved with their subsequent analysis. While doing fieldwork, I recognized the surfer and ethnographer together made a unique and politically powerful research team because of the type of information one methodology produces that the other does not. I experimented, routinely asking myself how that information informs one another.

of a recreational space.[7] My claims acknowledge the advantages and limitations of the citizen science data-gathering model that I used for this specific initiative.

Undergirding my critique of the initiative is a view of citizen science as a neoliberal concession to a neoliberal framework, dependent on the notion of the individual and individual responsibility.[8] That is, the responsibility rests more so on the individual rather than federal, state and local agencies to help fund good science and protect communities from environmental catastrophes including SLR. Here, I am sympathetic with the fates of scientists, who must forge ahead with conducting projects on a global scale without the necessary funding to do so without volunteer labor. The ethos of individual responsibility that characterizes many citizen science projects, Urban Tides included, therefore, points to the challenges that arise as local and federal funding for doing "good science" is slashed or disappears completely (Tucker & Taylor, 2004/2005, 29). Citizen science in Urban Tides' case substituted as a feasible alternative for expensive, expert-driven data collection. This state reflected the neoliberal political economy in which citizen science typically operates. It is dependent on unpaid labor and self-motivated behavior. The following section goes into deeper detail about Urban Tides' objectives.

Urban Tides Objectives

In California, Urban Tides regionalizes and localizes efforts to conserve, preserve, and protect coastal communities.[9] The broader vision entails actualizing governmental "strategies that will help the region adapt to the future impacts of SLR." The 2016 Urban Tides initiative encompassed four objectives. The first objective involved building a repository of public photos for scientists "to visualize current flooding risks at coastal locations in Southern California." These photos were synced from data bitches' smartphones and stored in Liquid's cloud-based storage system, which aimed to digitize and expand the movement of data that scientists (and other registered users) could utilize to determine processes such as spatial patterns. Urban

[7] To strengthen the transdisciplinary connections with ethnography and surfing, I had cross-fertilized my own ethnographic and surfer's knowledge into the sites where I chose to gather data. I created a small data set for Urban Tides specifically related to three of Trestles' surf breaks: Cottons, Lowers and Middles (located nearly a mile south of the Panhe Trail). I did not take photos of SLR at Church(es) or Uppers for Urban Tides.

[8] Sociologist Aya H. Kimura argues that neoliberalism's impact on contemporary models of citizen science signals an intricate bond between a "framing of politics as negative" and emotion-driven concerns for the future of the environment (2016, 155). Accountability, therefore, shifts onto individuals because fighting an environmental battle at the micro-level is easier than doing it at the corporate, governmental, or systemic one.

[9] Urban Tides emerged from the National Sea Grant College Program. This is a "partnership between universities and the federal government's National Oceanic and Atmospheric Administration." The USC Sea Grant's objectives derived from the passage of the National Sea Grant College Act in National 1966. As of 2025, the initiative was still in existence and open for data submissions through an online portal.

Tides' datasets were supposed to remain public, providing anyone with a Liquid account access to photos and recorded information about various beaches in the state. For this project, data bitches were deemed "contributors." Each record they uploaded to Liquid was asterisked with a reminder that they "[...] irrevocably consent that it may be used for educational, research, outreach and promotional purposes, in any medium, in perpetuity."[10] The implications of this caveat draw attention to the power relations inscribed in producing photographic knowledge about SLR.

Urban Tides' second objective is to "ground truth and calibrate scientific models that project flooding and erosion due to future SLR."[11] Oceanographers and meteorologists are trained to use technologies such as the Coastal Storm Modeling System for Southern California (CoSMoS) to help Urban Tides project SLR by 2050. They draw from CoSMoS to predict and prepare for environmental disasters (specifically coastal erosion and flooding) and the impacts of climate change. Oceanographer and meteorologist Andrea (Andy) O'Neill of the United States Geological Survey was instrumental in assessing the usability of my Trestles' SLR data and analyzing the citizen science data from the Urban Tides initiative. A series of features distinguish CoSMoS from other modeling systems that assess the impact of coastal hazards on California's coast.[12]

Determining and mitigating risk as a form of community resilience from coastal hazards is central to the initiative's third and fourth objectives. In this call for stakeholder collaboration, USC Sea Grant symbolizes bi-directional learning and knowledge-making among local communities, municipalities, city officials, scientists, and data bitches. Data, including that from Urban Tides' participants, is shared, discussed, and contextualized to increase ocean and climate literacy at the local levels. In the past, outreach efforts have included activities such as: educational beach walks; suggested lesson plans about king tides for elementary students; and presentations from public officials about SLR guidance and adaptation strategies. Workshops, webinar series, stakeholder meetings, and participation surveys are additional ways that USC Sea Grant has collaborated with local and state entities concerned about SLR. Researchers at the Scripps Institution of Oceanography (SIO), for instance, solicited feedback from Urban Tides citizen scientists via a pre- and post-participation survey in 2016 to better understand their motives for engaging in the initiative and

[10] Urban Tides Photo Database record.

[11] Urban Tides Community Science Initiative flier.

[12] O'Neill explained in a May 9, 2016, e-mail exchange: "The purpose of CoSMoS is to provide coastal flood hazard projections for future storms and SLR[.] We do that by explicitly simulating storm events with high-resolution numerical models, accounting for relevant physics and flooding contributions. Simulations are run for several storm (up to 100-year storm) and SLR scenarios [...] to provide a large ensemble of projections (over 40 scenario combinations). [...] All models have limitations, and there are many other projections and models out there with different strengths and weaknesses [....].

improve communication with scientists who use their photographs of beaches.[13] Ultimately, Urban Tides' objectives are grounded in the hope that data bitches' observations about how and where the coastline is changing "will help further collective dialogue about how we can adapt to rising sea levels."[14] In the following section, drawing from my own experience as a data bitch for Urban Tides, I explain the challenges that arise in trying to achieve the initiative's objectives.

Becoming an Urban Tides Data Bitch

In January 2016, I took the preliminary steps of volunteering for Urban Tides by attending field training at SIO in La Jolla, California. Scientists usually require those interested in their projects or initiatives to attend some form of field training. This is one way they attempt to achieve uniform data collection while navigating the methodological challenges of citizen science. This mode of preparation is also how scientists and researchers "implement protocol for rigorous data collection and submission that will engage motivated individuals who may or may not be formally trained in natural history observation [...] or scientific investigation" (Dickinson & Bonney, 2012, 20). Field training was expected to enhance citizen scientists' data quality and facilitate education outreach efforts.

My initial interest in attending field training stemmed from surfing. As a surfer, I was disturbed by the visible impact of winter storms mixed with extreme high tides (i.e., king tides) at various popular surf breaks throughout Southern California. During January 2016, for example, the main parking lot near the central lifeguard tower of La Jolla Shores, a tourist hotspot between the months of June and September, flooded after particularly intense winter-storm activity. Additional flooding occurred near a surf break named Trails in the southern zone of San Onofre State Beach, wiping out ten parking spaces that visitors and surfers frequently used to access the ocean. Several items were on the agenda for Urban Tides training, including an early morning walk from the parking northward along the beach to the SIO Pier during a king tide; a tutorial on how to photograph tidal lines using my smartphone; and two presentations about the scientific fundamentals of El Niño. With this training, I left with the hope that if I took photos of tidal lines at San Onofre State Beach, perhaps they would help scientists map SLR at Trestles and expand plans for community resilience to the City of San Clemente. Long term, the research could lend insight into different SLR imaginings of anthropogenic activity within and around the Anthropocene Ocean. This lofty imagining factored into my motivations for attending training (Fig. 11.1).

---

[13] See Astrid Hsu's "Walking North: Lessons in Citizen Science" (March 15, Hsu 2016). Accessed January 26, 2018. According to survey results: 45 percent of respondents cited "Project updates from scientists on how they are using the photos" as the most important component for encouraging participation in the Urban Tides initiative.

[14] Urban Tides Community Science Initiative flier.

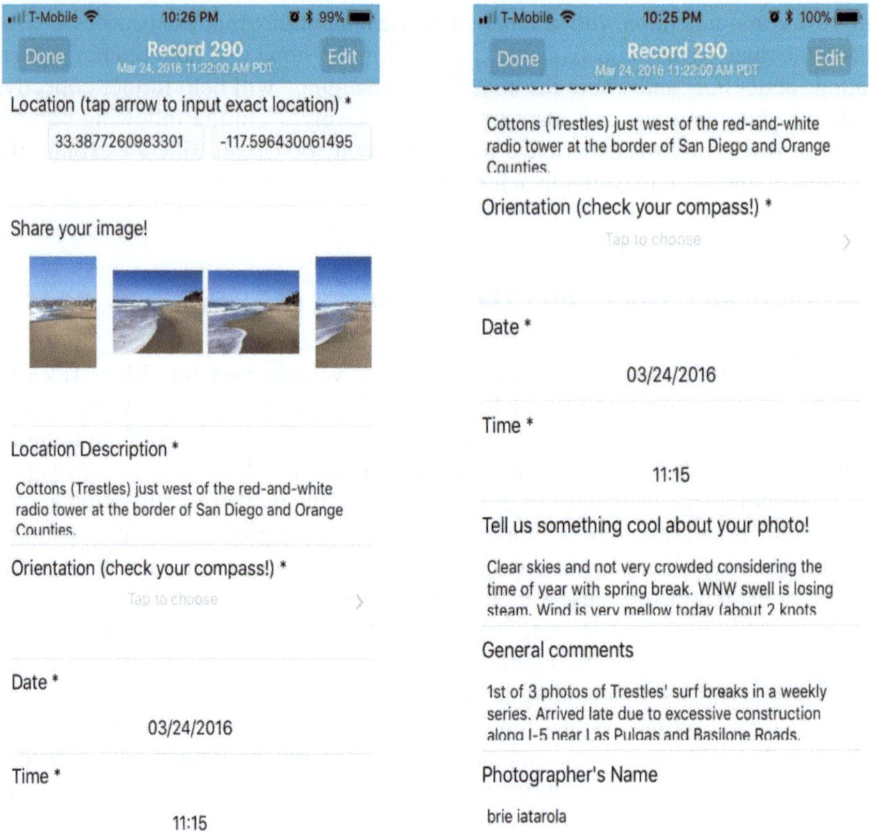

Fig. 11.1 An example of a completed urban tides record for March 24, 2016

I officially began taking photos for Urban Tides on January 10, 2016, amassing a total of 121 records by January 16, 2017. I describe here what I did as a data bitch, drawing upon different ways of doing citizen science and knowing the site, which in this case is technically Trestles, located within San Onofre State Beach. To become an Urban Tides contributor, I established a username and password, then downloaded the Liquid app on my iPhone 6s, unconcerned about the pivotal role that software updates or syncing would play in Liquid's database management.[15] With this feature, the Urban Tides data set automatically loaded once this information was generated, the app opened, and login was complete. I created a new digital record for each photo session and input required information, which included: location (auto-populated latitude and longitude with the tap of a small arrow); location description;

[15] In September 2016, no similar applications were available for download for other smartphones (e.g. Android). Desktop computers sufficed for uploading digital photos; this method took a few more steps to complete.

and compass orientation. The fields for date and time were auto-populated.[16] This especially mattered to GIS experts, oceanographers, and geographers who were using the Urban Tides data sets to model short- and long-term impacts from climate change along California's coast. Each photo included space to note "something cool" about the site and general notes that may be useful.

Each trip by car to San Onofre State Beach took between 35 and 55 min each way, depending on the traffic season.[17] I did not live inland, and my timeframe to leave was between 7:45 a.m. and 5:15 p.m., which made the commute generally fast and uninterrupted. The drawback was that my timeframe for collecting data always fell within traditional business hours. In March 2016 (an active month), I made five trips to Trestles, three of which I briefly describe in this chapter.[18] My ethnographic observations and on-the-ground knowledge as a surfer are meant to show how eco-ethnography and citizen science inform each other as methods of data collection and analysis, both of which allow me to see and document things that I might not have otherwise. Each month I set aside at least four hours to put on my data bitch hat, drive to San Onofre State Beach, and conduct fieldwork, which included taking photos of SLR.

For fieldwork on Friday, March 4, 2016, I departed La Jolla at 9:11 a.m. and parked off Cristianitos Road by 9:51 a.m. To get to the ocean, I usually took the Panhe Trail, even though this twenty-minute walk added time to my trip. It was an intentional choice, influenced by the victory of the Save Trestles and anti-241 Toll Road campaigns. When I arrived at the beach, I engaged in a ritual of picture-taking protocol after I decided which of the five surf breaks within Trestles that I would photograph first based on tide times. The Urban Tides initiative's directions stated the following:

> Take a few moments to watch the water. Find the highest wet water line in the sand. Then, take two steps or strides landward from the water line.
>
> Take the photo facing parallel to the shoreline. Include some sort of structure or feature in your picture, such as a pier, jetty, breakwater or dock, for perspective. This will help scientists better identify the water line.[19]

The first low at Middles was dropping to − 0.1 feet at 12:31 p.m.[20] That gave me 2.5 h to: walk south to the surf break; record beach-use activities; take notes about

---

[16] As Benjamin Zuckerberg and Kevin McGarigal explain: "Using data that are spatially explicit (i.e., sites that have exact latitude and longitude) is necessary if the investigator plans to delineate landscapes" (2012, 122).

[17] Goffredo et al. argue: "Asking volunteers to travel at their own expense to specific sites […] carries the risk of making participation in the research project less attractive and so reducing the number of volunteers willing to participate" (2010, 2171).

[18] Trips to Trestles occurred on the following dates in 2016: 3/4; 3/10; 3/16; 3/24; and 3/28.

[19] Urban Tides Community Science Initiative flier.

[20] A "mixed semi-diurnal tide" means that both high tides and both low tides are of different heights.

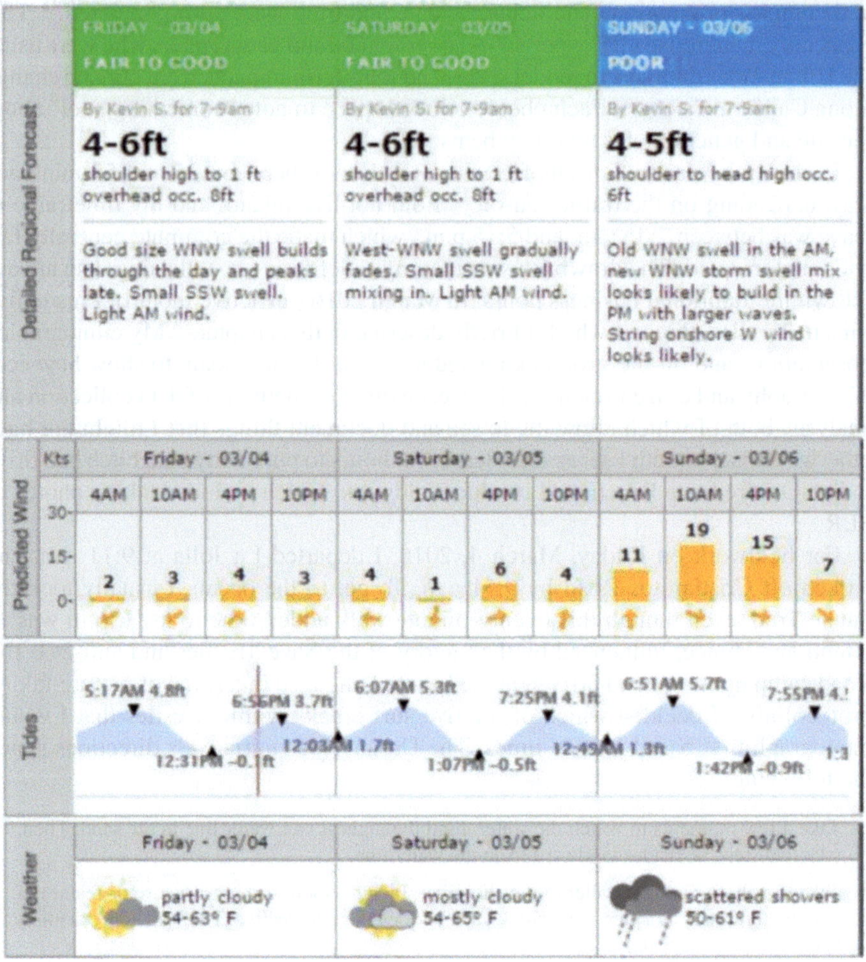

Fig. 11.2 A surf report from March 4–6, 2016, indicates wave size, predicted winds, tide information and weather. Image reproduction courtesy of Surfine.com

the surf conditions; compare my onsite observations to several surf reports; and walk back to my car (Fig. 11.2).[21]

Written inside my field notebook were notes for the three-day forecast for South Orange County from meteorologist and surf forecaster Schaler Perry of Surfline. On March 4, 2016, he reported: "Solid WNW building throughout the afternoon; favorable winds in the late morning and early afternoon. Dropping tide – 0.1 at 12:31 p.m. Fair to Good. 4–6 ft to 1 ft overhead occ. 9ft. Good size WNW swell builds through

[21] I elected to simplify my data collection efforts by using only one surf report from Surfline, which provided a representative sample of all surf breaks in South Orange County. My decision acknowledges methodological issues.

Fig. 11.3 Northbound traffic stalls at the San Clemente Border Patrol Station for U.S. Customs and Border Protection

the day and peaks late. Small SSW swell. Light AM wind." To translate: For the next few days, a swell from the west-northwest mixed with swell originating from the south-southwest was increasing in size. I did not know which observations technically qualified as "cool" per Urban Tides' criteria. Did surf assessments, crowd size, or my thoughts on race and gender matter to scientists mapping and modeling hypothetical sea level scenarios? Probably not, but through my data-bitch lens, they were important for explaining how space is produced at Trestles. Instead of submitting detailed write-ups through the Liquid app, I used a pencil to write down thicker descriptions in my field notebook. If only I also had followed through and printed out hard copies of Urban Tides data forms as a backup in the event data-syncing issues arose online (Fig. 11.3).

Traffic to Trestles on Thursday, March 24, 2016, was terrible with spring break in full swing. I missed the high tide by 35 min. Under heavy traffic conditions, driving a total of nearly 2.5 h to take photos of tidal lines for only 2.5 h was environmentally stupid. I was complicit in a transportation system and willingly contributed to greenhouse gas emissions because my concern for SLR at Trestles' future took precedence. Did that make me a misguided data bitch? Another challenge arose based on Urban Tides' needs: Scientists using the data sets for the CoSMoS project wanted images taken either at high (preferred) or low tide. Such timing shrunk my window for taking photographs and became an obstacle during commutes with heavy traffic.

I followed Urban Tides' photograph protocol while facing south toward the decommissioned San Onofre Nuclear Generating Station (SONGS) atop bluffs less than two miles down the coast. The scene captured a pre-Fukushima period in which the threat of tsunamis seemed an afterthought in plans for coastal development. I tapped an icon to activate the app's camera; this feature enabled me to capture real-time, geo-specific tidal activity. I added my name as the photographer in Record No.

Fig. 11.4 Capturing tidal lines at middles on March 24, 2016

226 for Middles, which included three photos and notes about the direction of the swell, adapted from Surfline. Circling north toward Lowers, I repeated the Urban Tides' protocol at 12:17 p.m. and created Record No. 227. *The swell appears to be less steeply angled than the one that came through last week*, I added as "something interesting." *The rocks are a little more scattered at this break than at Middles. Slope of sand is less extreme* (Figs. 11.4 and 11.5).

On windy Wednesday, March 16, 2016, I left La Jolla at 11 a.m., too late to reach Cottons in time for the 11:51 a.m. zero-foot low tide. I took the pictures anyway and trekked south to Lowers. Surfline forecaster Perry described the day's wind as "light to locally moderate offshore early, trending onshore through the morning. Light westerly flow 3–8 knots picks up in the afternoon, may lighten before dark."[22] Nope, this wind came from the northeast, and Surfline's projections were slightly weak. It was blowing harder than 8 knots, according to the wind tunnel in my ears (unscientific yet embodied surfer knowledge that was accurate enough to contradict the scientific surf report). I turned to Surfline's high-resolution wind-modeling system and wind barb tool for insight on how to think about these details because I knew wind combined with SLR contributed to coastal erosion. Once I reached Lowers, I pulled out my notebook, scrawling the *wind is much milder here and protected by the sand slope. Sand went from pebbly to hard-packed and firm (not quite dry, though).*

[22] Surfline utilizes its own model to produce high-resolution wind forecasts and advertises it is more sophisticated at capturing "small scale phenomena such as the sea breeze circulation and winds that interact with the terrain" than the Global Forecast System.

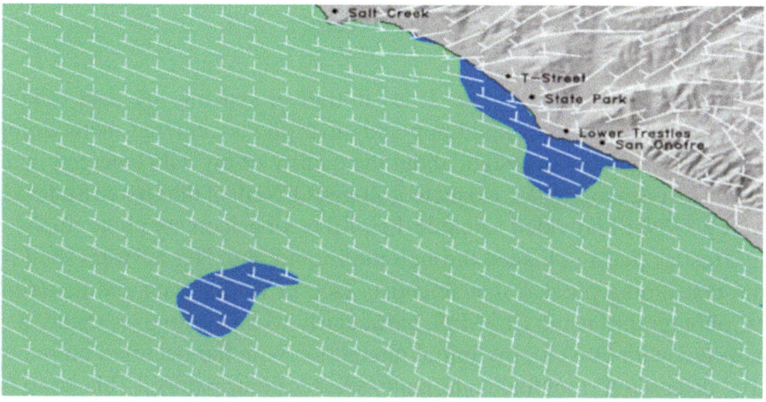

Fig. 11.5 A snapshot of Surfline's high-resolution wind forecast for 11 a.m. on March 15, 2017. Winds from the west-north-west were expected to blow between 6 and 10 knots heading toward San Onofre State Beach

Extreme slope from last few weeks seems to have leveled a bit. Much more crowded due to spring break most likely. [...] General bathing pool for the seagulls has shifted way southward. They're absent today. Upon reaching Cottons, I continued writing:

> *Wind indeed is coming from the WNW. I don't believe the arrows are correctly in place for Wednesday's three-day forecast -- at least at the start, but this is accurate. Yes to the WNW wind right now. It's the 5 p.m. arrow (arrow is pointing SE and wind is coming from the NNE, but the offshore swell heights and wind chart do not match this image at all. Must be some kind of error. Maybe it's hopeful wind versus predicted wind, and the hopeful winds take precedence for today? [...] Fuck arrows. I can feel the elements, which makes me understand their power better than any wind model or diagram.*

Photographing SLR at Trestles that day was uneventful, and I felt inklings of burnout and doubt. Would I have joined Urban Tides had the initiative not intersected with my research? Were my photos and observations useful or useless for scientists? I had no precise example for comparison. By the time I finished field work in January 2017, over the course of a year I had taken approximately 40 images per surf break and 30 pages of handwritten notes. The technological point of the Liquid app was to move away from traditional paper-based methods for data collection, yet my handwritten notes now served as a personal archive of memories and qualitative environmental data about SLR at Trestles. The following section assesses a set of limitations when the design of a citizen science initiative calls on non-scientists and surfers to use their smartphones and an app to photograph SLR.

Limitations of Apps for Citizen Science

In my experience with the Urban Tides initiative, syncing my data reliably was a notable limitation of the Liquid app and smartphone-centered methodology. On August 4, 2017, additional insight about the quality of my data emerged in a phone interview with oceanographer and meteorologist O'Neill of the United States Geological Survey. O'Neill provided valuable, helpful feedback on my photographs of tidal lines at Trestles that I uploaded through the Liquid app. Her answers shed light on the time-consuming process that scientists face in vetting data and modeling data bitches' photographs using CoSMoS, particularly when syncing issues arise. Before this interview with O'Neill, my initial interaction with her occurred online through Liquid's "discussion" tool for "community science contributors." This was a productive feature. I received an email alert that O'Neill had posted a comment regarding my descriptions of sand activity in photos of Middles. "Good context," she wrote, along with a request for me to move closer to the shore so that geographers had a better perspective of the tidal lines. Her comment opened a window for communication between the data bitch and scientist. I re-calibrated the following weeks, still unsure whether subsequent photos were usable, so I asked for confirmation through the discussion tool. A message from USC Sea Grant coordinators via email confirmed they were. The pace of my research and photo sessions remained steady. Eight months later, O'Neill and I conversed over the phone and a shared desktop program about the photos that I had uploaded through the Liquid app. She pulled two digital images from the Urban Tides dataset for comparison (Fig. 11.6).

The first photo was of Lowers at an approximate 1.49-foot low tide at 1:49 p.m. on October 26, 2016. O'Neill needed clarification on my location because she was unfamiliar with San Onofre State Beach. She noted in her email that location accuracy was USGS' biggest challenge in using the Urban Tides database.[23] O'Neill's cursor circled around the stones and pockets of water as she commented that it looked as though I were standing in the ocean, not quite far enough ahead of the tidal line. Furthermore, when mapped, the data point appeared too far out in the sea, which was a problem.[24] "Specifically, when you know all the conditions, then you can link similar conditions. These are all the high tides; these are the points," she said, switching to a mapped image where photos from Lowers were taken. Data points close together represented groups of similar tidal conditions. This helped give oceanographers more

[23] O'Neill added: "There are lots of pieces of geographic information that get tagged to photos (not only where the photo was taken, but sometimes where the photo was uploaded), and all those get uploaded to the site. So we sometimes have to do some detective work to use the correct coordinates. Beyond that, we also want to know how precise and accurate those photos are—images are taken from several different types of phones and cameras, and everyone's photos (no matter how carefully we craft guidance) are taken with slightly different perspectives[.].

[24] In our email interview, O'Neill noted that many citizen scientists took "really interesting photos of flooding impacts (of locations [she] is not familiar with) that [scientists] can't use—the coordinates are bizarrely for locations out in the middle of the Pacific Ocean." She continued by asking: "Where were those photos taken? What are the conditions at the beach? Did the citizen scientist upload from a ship?".

Fig. 11.6 A series of plotted coordinates from my urban tides photos of lowers. Mapped image courtesy of Andy O'Neill and USGS

confidence in what the geography was telling them, and how it was changing with other conditions, including extreme high and low tides. She wanted to know the story behind my photo. "That's Lowers at low tide," I said, adding that for each record I (almost) always included the information in "Tell us something cool about your photo!" There was a pause, followed by, "I see. That provides a lot more clarification on readings from low-tide photos."

I asked if my photos were useless, offering that SIO and Urban Tides indicated in field training that scientists would "welcome photos from any location, day and time; high tide, low tide, king tide."[25] O'Neill responded that although the scientific agency could "definitely use" my low-tide photos, they were "harder to validate against [...] selections not immediately relevant to [the] first order." She was using my high-tide photos to ground truth in radar data that USGS had. "Remote-sensing imagery can be quite powerful," she said, "but without concrete data on the ground to calibrate what I'm seeing from the air, sometimes it's hard to have confidence in what the geography is telling." For the CoSMoS project, high-tide photos were preferred—that is, the highest wet water line at high tide.

Two specific characteristics made a data bitch's photos usable: "good coordinates" and strategic location. Effective photos were positioned "along the beach with some sort of discerning feature or landmark in the view," O'Neill said. She also explained that usable photos helped USGS scientists and oceanographers "identify any behavior in the area (such as areas of extreme run-up or overtopping) that may be

[25] This was also the exact wording from the September 2016 Urban Tides flier.

important." Supplemental location information increased the agency's confidence in the photo's narrative, O'Neill added. For quality-control purposes, oceanographers combed through each of the database's uploaded photos, plotting their locations. "If the photo aligns with the imagery of the site and surrounding landmarks (that is, there isn't a mismatch between the coordinates and what's shown in the photo)," O'Neill said, "it goes into our own database for further use."[26] Additionally, mid-tide photos, of which O'Neill noted many data bitches submitted, had no value in this specific project. I winced upon realizing that most of my March 2016 photos were taken at low tide because the timing was conducive to my schedule. It would have been too time-consuming to comb through the remaining 121 uploaded records to compare the number of my high-versus low-tide photos. *So much for my useful contributions*, I thought, feeling like a dedicated yet failed data bitch—at least in terms of helping USGS and contributing usable photos for the CoSMoS 3.0 mapped results. O'Neill cited lack of Wi-Fi access and spotty cellular service at certain beaches as syncing obstacles. Inaccurate coordinates and locations for the bulk of my photos did not match the tidal line positioning, a problem solved only by human intervention and labor.

The situation exposed a major weakness in Liquid's syncing capabilities. The data-management platform was designed to help facilitate communication between scientists and Urban Tides volunteers, which it did to an extent. Each record was a data space for data bitches to communicate near-real-time environmental details to inform scientists' projections about SLR. But I learned of the syncing issue on my own on Thursday, June 2, 2016, when I logged in to view the Urban Tides dataset. None of my photos from April or early May were there. I e-mailed Urban Tides' communication manager to express my concern. She investigated the issue and discovered my photos were "unsynced." I confirmed I was using my iPhone 6s (with reliable cellular service) to upload them onsite, but currently the number 0 was listed under each record. I was instructed to log into the app, go to the records screen, and drag it down with my finger to force-sync my records to the larger dataset. If a checkmark did not appear after doing so, then I had to open and tap "done" to sync each record. The first suggestion did not work, so I tried the second approach as Liquid database managers attempted to sync my missing photos because my case was unique system wide. By July 25, 2016, the app appeared to be in working order, and a pending update was expected to fix additional bugs. My records no longer were labeled with 0, and my photos from April and May 2016 finally synced.

Until my discussion with O'Neill, I had assumed the syncing situation was resolved, realizing it was not. O'Neill said the uncertainty and possible error in the location underscored why some scientists hesitated to use data bitches' data in similar projects. In the Urban Tides case, nothing matched correctly with my photos, and there was no survey-grade equipment to rectify a possible error in the location. The results of my participation as a data bitch for Urban Tides revealed a challenge

[26] O'Neill interview on Friday, August 4, 2017. Liquid's dataset retains all the original photos; only the original user/photographer has the authority to delete from the public dataset.

in using an app dependent on syncing SLR data to a cloud-based storage system.[27] In many of my photos, it was difficult for O'Neill to pick out geographically where the water line was without linking a specific coordinate to what was happening near it. Cleaning up the dataset was laborious, and USGS had a strict timeline. Projections were already out for Southern California, O'Neill noted. For administers of future citizen science initiatives that encourage surfers and non-scientists to help collect SLR data when app-based technologies are central to the project's design, the following section offers three best practices. One best practice addresses possible modalities for facilitating in-depth training; the second considers the utility of contributors' paper-based systems as a backup for recording qualitative observations. The third best practice acknowledges the value of opening and maintaining lines of consistent communication between scientists, project administers, and data bitches. These communication lines must be nourished throughout the duration of the project should syncing issues arise between the app and databases that use a cloud for storage.

Best Practices for Administers of SLR Citizen Science Projects

The first best practice acknowledges it is imperative that administers provide multiple modalities for training citizen scientists before and during the data-collection process. Limited funding was a predicament that USC Sea Grant faced. Both O'Neill and I agreed that more in-depth training for how to take usable photos could yield better photographic data from data bitches. "As we get more people trained in how the data in their photos is used, then I think that's really powerful," O'Neill added. "They say, 'Oh, I see how they're using this. [...] This is potentially very powerful for lots of other science projects in risk analyses." If USC Sea Grant had the resources for this initiative, a small team of professionally trained or highly experienced science photographers would have been the ideal group of data collectors (Goffredo et al., 2010, 2172). The economic fallout of anti-science ideology over time indicates that financial support for environmental research was and remains unsteady, severely reduced, constantly threatened, or nonexistent.[28] As a result, individual citizens, nonprofit community groups, private start-ups, and financially stressed academic

---

[27] I agree with Bresnahan et al. (2022) in their assessment of the limitations of Smartfin when they explain that "[...] it would be desirable to have alternatives to cellular communication for data upload" (7).

[28] Kimura attributes this shift to "government budget cuts, particularly in what is often seen as unproductive areas such as environmental protection" (2016, 15). See: *Science* magazine's "What's in Trump's 2018 budget request for science?" (May 23, 2017); Mandy Sackett's "NOAA's Sea Grant Program on Trump's Chopping Block" in EcoWatch (April 18, 2017); Keith Matheny's "Trump budget would kill Sea Grant funds, end Great Lakes coastal research" in *USA Today* (March 8, 2017); "Lawmakers aim to save Sea Grant: Federal budget cuts threaten OSU program" in the *Newport News Times* (March 2017).

institutions end up shouldering the responsibility for training to complete environmental tasks that were formally under the domain of or funded by federal and local governments.

If this trend continues, future citizen science project designs should account for training modalities that are available online, in-person, in writing, and pre-recorded as video tutorials. Online modalities could defray additional in-person training costs while facilitating more in-depth guidelines and scientific reasoning about the preferred tidal lines for SLR photos. With ongoing training, citizen scientists could develop a more nuanced understanding of ideal photographic conditions, proper camera angles, and optimal locations, times, and coordinates for specific geographic contexts. Expanding the notion of collaboration in community science initiatives to account for students from community colleges and universities might widen and diversify the participation pool, although lack of access to reliable transportation remains a barrier to participation. Depending on the network of collaborators, academic course requirements could include teaming up students with more stakeholders to produce lower-stakes deliverables.

I understand my next best practice may seem outdated and impractical, but a personal paper-based system for recording qualitative observations in handwritten field notes remained methodologically necessary. Contrary to Shidler's hopes for Liquid, my Urban Tides data initially stayed "siloed in the sciences" because of this app-based system (Smith, 2015). Despite the critiques about the overall inefficiency and lack of scalability with a traditional paper-based system for SLR data collection, the cyber-structural weaknesses in Liquid's data-management system resulted in public data extinction. The issues made my traditional paper-based method a reliable but inconvenient alternative for remembering and contextualizing my coastal observations. Even with routine software updates, Liquid's real-time functions typically did not work as intended for my data. I subsequently lost all digital photos taken with the Liquid app's camera feature because they were uploaded and stored on a private server. My waterproof journal filled with dated, time-stamped, handwritten fieldwork notes from Trestles cemented my commitment to my traditional handwriting practices. It also animated my interest in re-imagining the archival process for storing and safeguarding citizen scientists' SLR photography and fieldwork notes.

The third best practice considers the function of science communication. I was grateful for the opportunities to communicate with Urban Tides administers and scientists who were involved in the initiative, including O'Neill from USGS. My feedback as a data bitch facilitated a tri-directional exchange of technological knowledge with Urban Tides administers when I helped identify a software bug in the Liquid app. Despite no formal training as a natural scientist and a small repertoire of technical vocabulary about SLR, eleven years of surfing already had boosted my ocean literacy. I knew surf reports and swell forecasts often mismatched actual wave conditions at my local surf breaks. My embodied knowledge of the environment as a surfer often interfered with or substituted for these inadequate and unreliable technological tools. Converting key details from my handwritten field notes and Perry's surf reports into "cool" eco-ethnographic observations for the Liquid app served a purpose for my own record-keeping methodology as a surf researcher. Whereas storm-modeling systems such as CoSMoS projected an aerial view of what SLR potentially would look like by 2050, my ground-level observations corroborated the impacts of El Niño, coastal flooding, and king tides throughout 2016. Downloading my coordinates data was the only power I had over their future use. Liquid's shuttering in June 2018 nearly rendered extinct everything that I had contributed for public knowledge throughout the year for Urban Tides. The communicative efforts and feedback from O'Neill and Urban Tides administers kept my faith in the initiative from fizzling. Their willingness to discuss the functionality of the Liquid app and dissect some of my photos from Trestles helped me (in)formally understand Urban Tides' educational value and coastal literacy dimensions.

In closing, although my personal experience volunteering for Urban Tides was different than I was expecting, I refrain from dismissing the entire enterprise of citizen science. It comes with its own set of challenges of how scientists and administers can extend their citizen science projects to better represent the communities impacted by SLR but unable (or unwilling) to participate in resilience-focused initiatives. Promising citizen science endeavors such as the Smartfin project represent the benefits of a surfer-as-participant approach to aiding in the scientific process (Bresnahan et al., 2022). Despite the inherent privacy risks, keeping all data "freely available at the open data portal sites" (Bresnahan et al., 2022) is key for future citizen science projects that seek to democratize scientific knowledge. The underlying socioeconomic and geographic inequalities in Urban Tides' citizen science model called into question the deliberative foundation upon which citizen science purportedly rests; however, it is never too late to recalibrate. The core objective of many community science projects is to promote a collective exchange of knowledge—environmental, scientific, cultural, technological, and otherwise (Dickinson & Bonney, 2012; Cooper et al., 2012; Goffredo et al., 2010; Kasymova & Gaynor, 2014). If an app and smartphone are involved, they must perform as expected. When they do not, it is of paramount importance that traditional lines of communication among project administers, scientists, and non-scientists are open, transparent, and generative. As for the future of the Anthropocene Ocean, citizen science projects that involve surfers as data bitches who care about the fate of vulnerable surf breaks will swell in size because of SLR.

References

Bresnahan, P., Cyronak, T., Brewin, R. J., Andersson, A., Wirth, T., Martz, T., Courtney, T., Hui, N., Kastner, R., Stern, A., & McGrain, T. A. (2022). A high-tech, low-cost, internet of things surfboard fin for coastal citizen science, outreach, and education. *Continental Shelf Research, 242*, 1–10. https://doi.org/10.1016/j.csr.2022.104748.

Cooper, C. B., Hochachka, W. M., & Dhondt, A. A. (2012). Citizen science: Public participation in environmental research. In J. Dickinson, R. Bonney (Eds.), *The opportunities and challenges of citizen science as a tool for ecological research* (pp. 99–101, 108–110). Cornell University Press, Comstock Publishing Associates.

Dickinson, J. L., & Bonney, R. (Eds.). (2012). *Citizen science: Public participation in environmental research.* "Why citizen science?" (pp. 7–11, 25–26). Cornell University Press, Comstock Publishing Associates.

Goffredo, S., Pensa, F., Neri, P., Orlandi, A., Gagliardi, M. S., Velardi, A., Piccinetti, C., & Zaccanti, F. (2010). Unite research with what citizens do for fun: 'recreational monitoring' of marine biodiversity. *Ecological Applications, 20*:8, 2170–2172, 2184.

Grace-McCaskey, C. A., & Iatarola, B., Manda, A. K., & Etheridge, J. R. (2019). Eco-ethnography and citizen science: Lessons from within. *Society & Natural Resources, 32*(10), 1123–1138.

Hsu, A. (2016, March 15). *Walking north: Lessons in citizen science.* USC Dornsife Guest Blog. Accessed on March 17, 2016 and May 5, 2025.

Kasymova, J., & Gaynor, T. S. (2014). Effective citizen participation in environmental issues: What can local governments learn? *State and Local Government Review, 46*(2), 14.

Kimura, A. H. (2016). *Radiation brain moms and citizen scientists: The gender politics of food contamination after Fukushima* (pp. 155–156). Duke University Press.

Liboiron, M. (2017, April). *Designing feminist technologies.* Workshop at UCSD.

National Oceanic and Atmospheric Administration. *Who we are.* seagrant.noaa.gov/WhoWeAre/SeaGrantHistory.aspx. Accessed on March 11, 2017.

O'Neill, A. (2017). E-mail interview. May 9.

O'Neill, A. (2017). Phone interview. August 4.

Pandya, R. E. (2012). A framework for engaging diverse communities in citizen science in the US. *Frontiers in Ecology and the Environment, 10*(6), 314–317.

Smith, C. B. *The power of liquid: A startup story. Cincinnati.com* (USA Today Network). (2015, September 24). www.cincinnati.com/story/news/2015/09/24/power-liquid-startup-story/72748176/. Accessed on August 4, 2017.

Surfline. (2014). *Upper trestles and lower trestles Southern California regional overview.* www.surfline.com/travel/index.cfm?id=2081#2950. Accessed 2014, March 1.

Surfline. (2007). San Diego County High Resolution Wind, 2007, March 15. www.surfline.com/surfdata/chart_viewer/?chart=106076id=2950zone=-8].

Tucker, C., & Taylor, D. (2004/2005). Good science: principles of community-based participatory research. *Race, Poverty & the Environment, 11*(2): 25, 27, 29.

Urban Tides Community Science Initiative. (2016). http://dornsife.usc.edu/uscseagrant/urban-tides-initiative/. Accessed on 2016, January.

USC Sea Grant. (2016, September). *Urban tides community science initiative: Capture the future of our urban ocean flier.*

Zuckerberg, B., & McGarigal, K. (2012). In J. L. Dickinson, & R. Bonney (Eds.), *Citizen science: Public participation in environmental research. Widening the circle of investigation: The interface between citizen science and landscape ecology* (p. 122). Cornell University Press, Comstock Publishing Associates.

Chapter 12
Drones, Phones, Boards, and Bills: How Law and Tech can Ensure Recreational Coastal Access Under a Changing Climate and Coastline

Jesse Reiblich

Introduction

What does the future hold for surfing and access to the coasts for recreational use in the Anthropocene? Proliferating technologies can both threaten privacy and provide opportunities to protect the environment and legal rights. That said, we need to evaluate what these technological advances and applications potentially mean for ensuring coastal access for recreational uses, like surfing as climate changes. On the one hand, improved satellite, drone and other monitoring technologies have all but destroyed the surfer's notion of secret spots. On the other, these technologies might be harnessed to safeguard the public's right to pursue coastal recreational opportunities, such as surfing. Complicating these potential upsides, is the fact that surfers are notoriously secretive and protective about the location of prized surf breaks. Even where they are known and publicized, surfers will sometimes band together to keep "non-locals" out. For instance, surf breaks like those in the historically exclusive Hollister Ranch development in Santa Barbara, California, have been safeguarded from outside visitors by various purposeful and de facto circumstances, including some that might violate California's Coastal Act (California Coastal Commission, 2014) and Hollister Ranch residents have fought tooth and nail to avoid allowing access to the surf breaks there to which they enjoy almost exclusive use.

Coastal recreational uses, such as surfing, rely on access to the coastline. As seas rise and erosion threatens beaches (Sweet et al., 2022), access is threatened (Caldwell & Segall, 2007). Human interventions like development and privatization of the commons on and close to coastlines also threaten coastal access. Specifically, coastal private property owners may seek to safeguard their property rights to the detriment

J. Reiblich (✉)
Assistant Professor, Department of Marine Affairs, University of Rhode Island, Kingston, USA
e-mail: jesse.reiblich@uri.edu

L. Mach (ed.), *The Future of Surfing in the Anthropocene*,
https://doi.org/10.1007/978-3-032-09377-6_12

of the public by armoring their properties, which can block or diminish coastal access as well as degrade coastal ecosystems and cause or exacerbate erosion. Compounding this issue further is the shifting nature of beaches, which makes determining where public property ends and where private owners may exclude all others difficult. But technology may offer the public, as well as coastal resource managers, tools to ensure public access to these traditionally common spaces. For example, drones and cameras can track the shoreline, and therefore where the public can legally access laterally along the shore (Adade et al., 2021; Holman et al., 1993). Other tools, such as mobile phone applications using augmented reality technologies and ones cataloging beach access points can help keep the public informed about their perpendicular access rights and potentially counteract nefarious efforts to block coastal access, such as illegal signage or physical barriers. Finally, laws and policies can be enacted or amended that better protect access despite rising seas and erosion, by using adaptive management and incorporating technological monitoring. This chapter introduces these concepts with an eye toward providing coastal resource managers a robust toolkit of options for ensuring public coastal access, despite climate change and other pressures on coastal resources.

Defining Coastal Access and the Shore in the Recreational Context

Coastal Access

"Access" and indeed "coastal access" can be defined in a variety of ways (Alterman & Pellach, 2022; Ribot & Peluso, 2003). Even amongst recreational users of the coasts "coastal access" can mean very different things. For a surfer, coastal access means at a minimum a footpath or some other accessway perpendicular to the coastline, which affords them access to enter the water close enough to paddle to where waves break (i.e. to the surf break). Other recreational users, like anglers, might require the ability to access the beach and traverse it parallel to the shore, and to remain there while fishing, or to remain to do other recreational activities from the beach. Additionally, even a surfer might want to access laterally across the shore to access other surf breaks, or perhaps to return to their perpendicular footpath after they finish surfing. Accordingly, for the purposes of this chapter "coastal access" means either the opportunity to perpendicularly or laterally access the shore or coast to perform recreational activities, such as surfing.

The Shore

Similarly, "the shore" (also known as "the coast") is legally defined differently depending on jurisdiction. Colloquially, the shore is synonymous with the beach and other littoral and riparian locations where waterways meet land. While the shore and coast may seem obvious or easily defined, legal and practical issues can emerge when they are not properly defined, either by lawmakers or by judges interpreting the law. For instance, in Rhode Island, the state supreme court was left to define the meaning of "the shore" in that state's constitution because it was undefined in state law (State v. Ibbison). Subsequent efforts to clarify the meaning of this term in legislation have been met with lawsuits (Amaral, 2023). Regardless, efforts to define the shore, in Rhode Island and elsewhere, seek to more definitively define the shore for the public as well as private coastal property owners.

For the purposes of recreational access and this chapter specifically, the shore includes anywhere accessible for coastal recreational pursuits, typically between the low tide mark and any vegetation upland (e.g. Texas Open Beaches Act). However, the legality of accessing the shore varies by jurisdiction. In some places, the legally accessible shore encompasses the entire wet sand beach and dry sand beach, up to the vegetation line or another similarly stable demarcation (e.g. Oregon and Hawaii in the United States). In other jurisdictions, like Massachusetts and Maine in the United States, access is only ensured for the wet sand part of the beach, and only for certain recreational pursuits. For instance, in these states both fishing and hunting are protected recreational uses under the public trust doctrine, and even SCUBA diving is recognized in Maine, but surfing is not necessarily permissible on these privatized shorelines (Reiblich & Reineman, 2018). Most other jurisdictions, in the United States and abroad, fall somewhere between these extremes. Further, even when the legal definition of where private property ends and public property begins, or alternatively, which private parts of the shore are nonetheless open to the public for certain uses under the public trust doctrine (e.g. fishing and hunting), actually drawing this line can be complex for various practical reasons. For example, because beaches naturally shift, even daily but certainly seasonally, the actual boundary line is typically an average of locations of high-water lines or similar (Borax Consol. v. City of Los Angeles). As such, any access issues identified and examined in this chapter should be considered within the local specifics of legally accessible shoreline boundaries for that jurisdiction.

Threats to Coastal Access

Climate Change

Perhaps the primary existential threat to coastal access is climate change. Specifically, of the threats discussed in this chapter, the effects of climate change have the most potential to limit or destroy coastal access. The effects of climate change on the coasts are primarily seen in sea level rise and the related effects from eroding shores. Seas are rising and are expected to rise between 0.95 and 3.61 feet by the end of the century (Intergovernmental Panel on Climate Change, 2019). However, some areas are experiencing these effects to a greater or lesser degree because of regional uplift, subsidence, and other local factors. While rising seas can affect coastal access points, and access generally, by inundating these places, erosion can affect coastal access by destroying parts of the shore that provide access to the coast (e.g. Beacons Beach in Encinitas, CA). Other more attenuated effects of climate change will affect coastal access as well. For example, hardened structures, built to respond and fight back against rising seas and erosion, can block access. Coastal squeeze, discussed in the next subsection, is similarly linked to the effects of climate change as it is combined with development of the coast.

Development and Privatization

Development and encroaching private property development also threatens coastal access. These phenomena exacerbate the effects of climate change and cause "coastal squeeze" along the coasts. Coastal squeeze is the disappearance of beaches and coastal features as seas rise, coastlines erode, and development stops the otherwise inevitable migration of the shore inland (Pontee, 2013). Coasts naturally shift and migrate, typically inland, but built infrastructure can block this natural migration, therefore causing coastal squeeze and reducing beach and coastal surface area. The disappearance of these coastal areas, such as beaches, likewise impacts and reduces coastal access as well.

Further, and even more troublingly, while coastal areas have traditionally been common areas, open to all, development and neoliberal policies have increasingly privatized these areas and closed them off to the public (Kahrl, 2018). For instance, a wealthy tech executive purchased the property upland of Martin's Beach in California. The new owner closed a previously open road that beachgoers have used to access Martin's Beach for decades (Balmer, 2018). Luckily for users of that beach, California's Coastal Act required the new owner to reopen the access road, at least temporarily, and to follow certain development procedures to close it permanently. But other states with less robust coastal access policies might not be similarly suited to be able to counter attacks like this one on access in their jurisdictions. Further, as populations and coastal development increase, and conflicting coastal uses likewise increase, conflicts over coastal places are likely to increase as well.

Other Threats

Coastal access is further threatened by other things, such as illegal signage (Ponting & Mach, 2025), policymaking that transfers coastal stewardship to private parties, as well as judicial opinions limiting the uses of the coast pursuant to the public trust doctrine. For example, in several jurisdictions coastal property owners have posted signs that claim incorrectly that access to the coast is not permitted at some public access points or that parking is illegal even where it is not. Perhaps the most egregious example of a coastal landowner blocking access in this way was when famous record executive, David Geffen, erected fake garage doors on the sides of his Malibu, California home to deter anyone from parking next to his home and accessing the public beach access points nearby (Abcarian, 2014). There are also examples of some towns removing signs that show where access points or beaches are located to keep crowds down and nonresidents from being able to access these beaches. Other less overt policies also deter and threaten access, such as disappearing public parking spaces and lots and the proliferation of beaches that charge money to access the beach or to park nearby (Thompson, 2015). Specifically, these efforts deter the disadvantaged and those with lesser means from accessing the coast. Similarly, efforts to stymie public transit to the coast or bold nefarious policy decisions to block busses and trains from extending routes to beaches to keep poorer or minority people from accessing them have also occurred (Kahrl, 2018).

How Technology and the Law Can Ensure Coastal Access

Monitoring

One way of ensuring coastal access is by monitoring where the coastline, and therefore the public part of the shore, is open to the public. Importantly, this dividing line depends on jurisdiction, with some states drawing the line at the mean high water mark or others at the mean low water line, with other variations in other jurisdictions still. Ensuring access by monitoring this line is perhaps most important where this line most favors private coastal landowners (i.e. where the line is at the mean low water mark). However, monitoring is an important way to protect coastal access everywhere, especially in the face of climate change and coastal squeeze, and as private property owners seek to retain their exclusive and exclusionary coastal rights, even as seas rise, and their properties potentially shrink.

Cameras

The most traditional technological tool discussed in this chapter that can help protect coastal access is fixed video camera monitoring mounted in specific coastal locations. Because state and local governments typically own and manage their sovereign submerged lands and various public beaches and other coastal public spaces, this option would potentially be the most appropriate if these entities seek to monitor their coasts, and relatedly coastal changes and other phenomena. For example, states and local governments could install fixed coastal cameras to track and monitor the shifting location of the coastline. However, an obvious drawback or challenge for this option will be that there would need to be many of them, and they would need to be placed in potentially precarious locations along the coast. Additionally, this challenge would be exacerbated, and potentially rendered impossible, in jurisdictions where the dividing line between public and private coastline is at the mean low water mark (e.g. Massachusetts and Maine). Video monitoring and studies of the coastline using video technology has occurred for several decades (Lippman & Holman, 1989). A primary motivation behind adopting this technology was because it is cheaper than satellite monitoring (Holman et al., 1993). Video imaging has been successfully used to identify the erosion rate of a pier on a beach (Pianca et al., 2015). More recent technology included networked camera monitoring techniques, such as the Argus system, which allows several cameras to provide monitoring in the form of a 180° view (Bruder et al., 2019). Alternatively, researchers have studied whether existing networks of surf cams can be used to monitor coastal variability and change (Bracs et al., 2016). However, because these cameras were installed to monitor the surf conditions rather than the shoreline, their applicability may be limited (Bracs et al., 2016). The drawbacks regarding using fixed camera monitoring to track coastal change, rising seas, and to potentially safeguard public coastal access are to some extent remedied by the other technologies discussed in this chapter.

Drones

Unmanned aerial vehicles, colloquially called "drones," can similarly be used to monitor the coasts when they are equipped with cameras. Drones are an attractive option for monitoring the coasts because one of them can cover a lot more ground than a fixed camera. Similarly, they can monitor many angles of an area and coastline instead of only from a fixed vantage point. They are also potentially preferable to fixed cameras because it can be more affordable to use drones instead of fixed cameras or using satellite technologies (Adade et al., 2021). Drones can provide highly detailed, high-resolution imaging at relatively low cost (Young, 2021). Drones are also relatively easy to use because they can be preprogrammed and operated remotely (Adade et al., 2021). One challenge of using drones is that their use can implicate privacy issues (Hicks, 2017). Additionally, they are subject to local, state and federal regulations regarding their use (U.S. Department of Transportation, 2023). There has been at least one review of case studies where drones have been used in coastal zone management applications (Adade et al., 2021). This review identified several examples, including how drones have been used to track coastal erosion (Pucino et al., 2021), map topographic and geomorphic changes at beaches (Chen et al., 2018), and to monitor slope stability of sea cliffs (Barlow et al., 2017). Drones have also helped facilitate long term morphological monitoring of beaches (Jaud et al., 2019), and to monitor coastal flooding (Addo et al., 2018). Studies have similarly examined whether off-the-shelf drones are feasible for monitoring the erosion and sediment of living shoreline projects over a two-year period (Young et al., 2021).

Specific concerns regarding using drones to monitor the natural environment include privacy, safety, and data security, among others (Sandbrook, 2015). Perhaps most obviously, drones have the potential of raising serious privacy concerns if they trespass into private property or if they purposefully or incidentally capture private things on camera. Further, drones can cause safety concerns for aviation if they interfere with flight paths or otherwise impact the safety of planes and other flying craft. These concerns have been lessened as drones become more advanced and more accepted by the public (Merkert & Bushell, 2020).

Citizen Science

Another way of monitoring the coastline would be to ask surfers and other beachgoers to upload digital photos of the coast as it appears using their mobile devices, typically a cellular phone equipped with a camera. Using crowdsourced photos to monitor the coasts is a kind of rudimentary citizen science. This method is attractive because it can provide up close photo documentation in ways drones and fixed cameras might not be able. Additionally, it obviates the need for a centralized drone team, presumably run by a local government or an academic institution or nonprofit, to purchase and operate drones. Instead, beachgoers can use their mobile phones that they already own. Surfers have been noted as having keen understanding and local knowledge of the shore and coastal environments (Reineman et al., 2017). An Australian citizen science project that tracks coastline changes now operates in twenty-one countries (Harley & Kinsela, 2022). Similar citizen science programs have been used to track marine debris locations (Marine Debris Tracker 2024) and to document king tides (Virginia Institute of Marine Science 2024). This method is also potentially superior to the others because it could go on as long as the phone application functions and people carry phones with them to the beach. Potential drawbacks are that the data would not be regimented or done in scientific intervals. Instead, the voluntary citizen scientists would contribute photos to the cause when they wanted. Citizen science's reliability tracking coastal erosion has been studied, including its relationship with drone monitoring (Pucino et al., 2021).

The drawbacks of citizen science primarily revolve around participation by the public. For instance, certain more popular locations would receive more attention and uploads of pictures than less visited or more remote locations. Similarly, a location might be more visited during certain seasons or perhaps more often when recreational conditions are better (e.g. surfers might visit beaches more often when there are waves than when conditions are flat). Citizen science efforts might also be less helpful than others on this list because they are limited by the vantage points available to beachgoers. This limit might skew the data uploaded to photographs that do not show the entire coastal picture in the same way the "bird's eye view" angle of a drone can. Regardless, citizen science is a helpful tool to understand coastal change, particularly to highlight changes that might not otherwise receive attention immediately. For more on the opportunities and challenges involved with conducting citizen science on coastal erosion and access, Brie Iatarola offers a detailed analysis of her experiences in chap. 11 of this volume.

Education and Increased Access to Information

Improving and increasing the quality and availability of education and information available regarding coastal access options and locations can also ensure access is safeguarded. One barrier to accessing the coast is lack of good, updated information about where the public can safely access the shore. One option to address this shortcoming has been for a jurisdiction to publish public access options, either in a book or similar media (California Coastal Commission, 2014). More recently, online and other digital computer viewers and websites have been built to convey this information more effectively and in formats that are easier to update. Mobile device applications can similarly provide this information using GPS and other technologies, such as augmented reality, to better convey where coastal access points are and other relevant information about them (i.e. amenities, access types, potential barriers to access, parking nearby).

Access Guides

Simple access guides are the most straightforward way to educate the public about their rights to the coasts. In the United States, coastal access guides are typically published by states. For instance, the state of California publishes a California Coastal Access Guide (California Coastal Commission, 2014). Rhode Island similarly has a guide called Public Access to the Rhode Island Coast (Allard Cox, 2004). In fact, most coastal states have some kind of coastal access guide, in physical book form or at least posted online in a printable format (e.g. Oregon and Florida's guides). One advantage of a physical, paginated, access guide is that it would work without power and without Wi-Fi, internet service, or cell phone service, as the other options in this section require. However, a drawback is that these options might be out of date and that updating them would be less instantaneous than electronic guides.

Mobile Applications

Additionally, education can come in the form of mobile device applications that document and list access points. Rhode Island's mobile version of its access guide, Discovering the Rhode Island Shore, is an interactive guide to that state's coast, and is updated based on user feedback (Rhode Island Coastal Resources Management Council, 2016). California's version of its coastal access resource is available as a mobile application called YourCoast (California Coastal Commission). Again, existing mobile coastal access applications are typically produced on a state-by-state basis. However, there are some more localized mobile applications. For instance, Malibu, California features a local application called Our Malibu Beaches (Our Malibu Beaches, 2021). This form of informative coastal access guides has several advantages. First, these guides are portable as mere applications added to mobile phones—portable devices that fit in pockets and that are typically always with us wherever we go. These application-based guides can also be updated centrally instead of needing to release updated print versions. Likewise, because mobile phones feature location sensing technology, these guides can be programmed to highlight the coastal access points closest to you at any given time. However, critics have pointed to potential errors in these mobile applications that have caused concerns, such as supposedly listing too much of coastal beach spaces as public, or negative impacts (e.g. littering) associated with increased use of otherwise lesser known coastal access areas (Abel, 2024).

Virtual and Augmented Reality

Relatedly, these educational resources can also harness virtual or augmented reality technologies. Virtual reality has been useful for explaining environmental issues, such as virtual visualizations of coral reefs under changing climate, and rising seas (Calil et al., 2021). While virtual reality uses wearable video and audio headsets to help a user visualize things that are not physically in front of the headset wearer, augmented reality showcases visualizations associated with global positioning system (GPS) locations on mobile devices in locations typically associated with whatever additional information or visualizations are being produced by the AR technology. Specifically, because of its use of GPS, and because it can be tied directly to specific, real, coastal locations, augmented reality technologies can be used for various coastal location applications. For instance, a user of AR technology can help a beachgoer identify exactly where coastal access locations are, without the need for signs or other designations. AR has been used as a tool to help communities visualize and communicate potential flooding risks (Haynes et al., 2018). At least one project has combined VR and AR to increase awareness for coastal communities like Long Beach and Santa Cruz, California (Calil et al., 2021). Additionally, it can safeguard against some coastal access issues, such as the removal of coastal access signage, or false or misleading signs aiming to dissuade coastal access users from accessing certain places—an issue that is becoming increasingly problematic as private landowners seek practical and effective methods to keep the public away from them and their property.

While these technologies can help increase awareness of sometimes hard to imagine phenomena like flooding and rising seas, and they can potentially be applied to help increase awareness of and perhaps protect coastal access, they have potential drawbacks as well. Incorrect information regarding coastal access points could cause increased crowds in some coastal locations or even encourage trespassing on private property. Similarly, backlash against efforts to educate the public about their coastal access rights could prompt unhappy private property owners to lobby their local decision makers and elected officials to close off or limit previously public access. For instance, disgruntled neighbours might advocate for restrictive parking policies on public roads close to these public access points.

Policies and Lawmaking

Finally, lawmaking and policymaking, including laws that feature technological and adaptive management aspects, can also protect and ensure coastal access. This section considers some of these policy measures that if implemented can help protect coastal access.

Rolling Easements

One legal strategy to ensure access to the coastline, particularly lateral access along the shoreline, is by enshrining rolling easements into law. While the shoreline boundary itself typically "rolls" with rising seas, and thus the property line between private landowners and the public automatically rolls inland as seas rise, enshrining a defined buffer above this natural movement is one way to ensure lateral access. These buffers are known as rolling easements, and in addition to ensuring increased lateral access, they also allow coastal features, such as wetlands and beaches, to naturally migrate inland (Titus, 2011). Laws have been implemented that act as rolling easements. For instance, Rhode Island's shoreline access law is a rolling easement (R.I. Gen. Laws § 46-23-26). More famously, Texas's Open Beaches Act uses rolling easements to ensure access as well (Tex. Nat. Res. Code Ann. § 61.011). Rolling easements are a resilient coastal management legal strategy because they naturally move inland as the shore erodes and recedes.

Prioritize Access as a Policy Goal

Another way to ensure access for recreational uses, is to draft laws and policies focused on maximizing access as a policy goal. California's Coastal Act seeks to maximize public access to the coast and also has as its goal maximizing recreational opportunities in the coastal zone (Cal. Pub. Res. Code § 30001.5). Several of the Act's provisions have requirements directly tied to this aim. For instance, development projects must include public access from the road to the shore (Cal. Pub. Res. Code § 30212). Additionally, low-cost visitor and recreational facilities are incentivized, including overnight accommodations (Cal. Pub. Res. Code § 30213). Another potentially effective way to prioritize the protection coastal access in policy is to put existing policies into practice. Specifically, jurisdictions that have existing rights of way programs can increase the budgets of these programs so that they can better carry out their missions. NOAA CZMA funding is one potential funding source for programs to increase their funding. Finally, states can empower one of their state agencies to improve access in their jurisdictions, as California has done with its State Coastal Conservancy (Cal. Pub. Res. Code § 31104.1).

Require that Shoreline Protection Measures Do Not Hinder Access

Laws and policies could also be designed to incentivize or require shoreline protection structures that do not interfere with coastal access. For instance, a Rhode Island law requires that any shoreline protection structure in that state not block access (650-20-00 R.I. Code R. § 1.3.6(B)(4) and R.I. Gen. Laws § 46-23-1(a)(1)). As a result, protective structures are sometimes designed with drawbridges or other design features that allow access despite of the structure being located in potentially shoreline blocking locations. Additionally, Virginia passed a law that requires landowners to use living shorelines instead of hardened structures unless they can prove living shorelines are infeasible (28.2-104.1(D) of the Code of Virginia). While this switch does not on its own ensure access, it can help avoid coastal squeeze scenarios typically caused by embracing hardened structures and therefore can ensure access in the face of rising seas or eroding beaches.

Other Options and Considerations

There are additional minor policies and law changes that could also help to ensure coastal access is preserved for recreation under the pressures discussed in this chapter. Stronger laws and policies could be implemented to address fake signs that attempt to deter recreational beachgoers from parking or accessing the coast using legal means. Additionally, recent laws limiting coastal access, like the one passed in Florida, could be rescinded (Flournoy et al., 2020). New coastal hazard and shoreline access and management laws could also consciously include adaptive scientific standards and strategies that are proactive rather than reactive. For example, laws could include language requiring managers to use the best available science in their decision making and to implement strategies like retreat when certain thresholds—sometimes called "triggers"—are met regarding areas flooding or experiencing other perceptible conditions, like erosion or erosion rate (Setter et al., 2023). Certain policy strategies could be adopted to gain more buy in and participation from private property owners, such as negotiated purchases of their flooding properties that do or will block coastal access, and limiting the liability of landowners who allow easements across their property for lateral and perpendicular access (Mukherji et al., 2024). Finally, more funding to identify, purchase, and improve coastal access points could similarly help to maintain and improve coastal access under a changing climate (New Jersey Department of Environmental Protection, 2024).

Conclusion

Coastal access is under threat from multiple pressures. As these pressures mount, there is a need now more than ever for innovative solutions to ensure access regardless of what the future holds. Improving technologies can help protect coastal access by clearly marking the public shoreline. Additionally, technology is making it easier than ever for the public to educate themselves about their rights to coastal access and where coastal access points are. Science-based and technological standards and innovations can likewise be coupled with progressive and adaptive laws and policies to better ensure access to the coast. To protect coastal access for recreational activities such as surfing, policymakers and grassroots organizers and activists can embrace the options discussed. While it might become harder to protect your favorite secret spot from being identified and found as technologies improve, these can nonetheless potentially be used to safeguard and even enhance access and the status of coastal spaces.

References

650-20-00 R.I. Code R. § 1.3.6(B)(4) (2022).

Abcarian, R. (2014). *Long overdue: Malibu elitists who impede public access now face fines*. Los Angeles Times, June 23. https://www.latimes.com/local/abcarian/la-me-ra-malibu-public-access-fines-20140623-column.html.

Abel, J. (2024). *Local homeowner asks beach access app to correct errors*. The Malibu Times, August 13. https://malibutimes.com/local-homeowner-asks-beach-access-app-to-correct-errors.

Adade, R., Aibinu, A. M., Ekumah, B., & Asaana, J. (2021). Unmanned Aerial Vehicle (UAV) applications in coastal zone management—A review. *Environmental Monitoring and Assessment, 193*(154). https://doi.org/10.1007/s10661-021-08949-8.

Addo, D. A., Abel, S., Ansah, R. K., & Nkrumah, I. (2018). Review on investigating the possibility of local hidden variable theories-quantum teleportation. *International Journal of Advanced Mathematical Sciences, 6*(1), 1–9. https://doi.org/10.14419/ijams.v6i1.8755

Allard Cox, M. (Ed.). (2004). *Public access to the Rhode Island coast* (p. 84). Rhode Island Sea Grant. http://www.crmc.ri.gov/publicaccess/ri_access_guide.pdf.

Alterman, R., & Pellach, C. (2022). Beach access, property rights, and social-distributive questions: A cross-National Legal Perspective of fifteen countries. *Sustainability, 14*(7), 4237. https://doi.org/10.3390/su14074237

Amaral, B. (2023). *Property owners' suit seeks to Block New R.I. Shore Access Law*. The Boston Globe, July 7. https://www.bostonglobe.com/2023/07/07/metro/property-owners-suit-seeks-block-new-ri-shore-access-law/.

Balmer, P. (2018). Martin's beach litigation and eroding public access rights to the California coast. *Ecology Law Quarterly, 45*(2), 427–438. https://www.jstor.org/stable/26568801

Barlow, J., Gilham, J., & Cofrã, I. I. (2017). Kinematic analysis of sea cliff stability using UAV photogrammetry. *International Journal of Remote Sensing, 38*(8–10), 2464–2479. https://doi.org/10.1080/01431161.2016.1275061

Borax Consol. v. City of Los Angeles, 296 U.S. 10, 56 S. Ct. 23, 80 L. Ed. 9 (1935).

Bracs, M. A., Turner, I. L., Splinter, K. D., Short, A. D., Lane, C., Davidson, M. A., Goodwin, I. D., Pritchard, T., & Cameron, D. (2016). Evaluation of opportunistic shoreline monitoring capability utilizing existing 'Surfcam' infrastructure. *Journal of Coastal Research, 32*(3), 542–554. https://doi.org/10.2112/JCOASTRES-D-14-00090.1

Bruder, B. L., Hathaway, K. J., Harrison, S., & Brodie, K. L. (2019). *Design and deployment of mini-argus systems for rapid coastal imaging*. ERDC/CHL CHETN-IV-114. Coastal and Hydraulics Laboratory (U.S.) Engineer Research and Development Center (U.S.). https://doi.org/10.21079/11681/31463.

Cal. Pub. Res. Code § 30001.5.

Cal. Pub. Res. Code § 31104.1.
Cal. Pub. Res. Code § 30212.
Cal. Pub. Res. Code § 30213.
Calil, J., Fauville, G., Queiroz, A. C. M., Leo, K. L., Newton, A. G., Mann, T. W.-W., Salvatore, P., & Bailenson, J. N. (2021). Using virtual reality in sea level rise planning and community engagement—An overview. *Water, 13*(9), 1142. https://doi.org/10.3390/w13091142
Caldwell, M., & Segall, C. H. (2007). No day at the beach: Sea level rise, ecosystem loss, and public access along the California coast. *Ecology Law Quarterly, 34*(2), 533–578. http://www.jstor.org/stable/24114634
California Coastal Commission. (2014). *California coastal access book* (7th ed.). University of California Press.
Chen, B., Yang, Y., Wen, H., Ruan, H., Zhou, Z., Luo, K., & Zhong, F. (2018). High-resolution monitoring of beach topography and its change using unmanned aerial vehicle imagery. *Ocean & Coastal Management, 160*, 103–116. https://doi.org/10.1016/j.ocecoaman.2018.04.007
Code of Virginia 28.2-104.1(D).
Flournoy, A., Ankersen, T. T., & Alvarenga, S. (2020). Recreational rights to the dry Sand Beach in Florida: Property, custom and controversy. *Ocean and Coastal Law Journal, 25*(1), 1–63. https://digitalcommons.mainelaw.maine.edu/oclj/vol25/iss1/2
Harley, M., & Kinsela, M. (2022). Coast snap: A global citizen science program to monitor changing coastlines. *Continental Shelf Research, 245*, no. 104796. https://doi.org/10.1016/j.csr.2022.104796.
Haynes, P., Hehl-Lange, S., & Lange, E. (2018). Mobile augmented reality for flood visualisation. *Environmental Modelling and Software, 109*, 380–389. https://doi.org/10.1016/j.envsoft.2018.05.012
Hicks, W. (2017). Comment: Maryland state drone law puts residents at risk of privacy intrusions from drone surveillance by law enforcement agencies. *University of Baltimore Law Forum, 47*(2), 130–145. http://scholarworks.law.ubalt.edu/lf/vol47/iss2/5
Holman, R. A., Sallenger, A. H., Lippmann, T. C., & Haines, J. W. (1993). The application of video image processing to the study of nearshore processes. *Oceanography, 6*(3), 78–85. http://www.jstor.org/stable/43924648
Intergovernmental Panel on Climate Change. (2019). *Special report on the ocean and cryosphere in a changing climate*. https://www.ipcc.ch/srocc/.
Jaud, M., Delacourt, C., Le Dantec, N., Allemand, P., Ammann, J., Grandjean, P., Nouaille, H., Prunier, C., Cuq, V., Augereau, E., et al. (2019). Diachronic UAV photogrammetry of a Sandy Beach in Brittany (France) for a long-term coastal observatory. *ISPRS International Journal of Geo-Information, 8*(6), 267. https://doi.org/10.3390/ijgi8060267
Kahrl, A. (2018). *Free the beaches: The story of ned coll and the battle for America's most exclusive shoreline*. Yale University Press.
Lippman, T., & Holman, R. (1989). Quantification of sand Bar morphology: A video technique based on wave dissipation. *Journal of Geophysical Research, 94*(C1), 995–1011. https://doi.org/10.1029/JC094iC01p00995
Marine Debris Tracker, effective November 23, 2024. https://debristracker.org/.
Merkert, R., & Bushell, J. (2020). Managing the drone revolution: A systematic literature review into the current use of airborne drones and future strategic directions for their effective control. *Journal of Air Transport Management, 89*(101929). https://doi.org/10.1016/j.jairtraman.2020.101929
Mukherji, A., Ke'Ziyah, W., Kayode, N. A., Millea, M., Scott, C., & Bella, S. (2024). Buyouts in the Carolinas: Pre & post buyout perspectives of public officials and community leaders. *International Journal of Disaster Risk Reduction, 113*, 104906. https://doi.org/10.1016/j.ijdrr.2024.104906
New Jersey Department of Environmental Protection. (2024). *Blue acres program*. https://dep.nj.gov/blueacres/.

Our Malibu Beaches. (2021). *The definitive guide to the public beaches of Malibu.* https://ourmalibubeaches.com/.

Pianca, C., Holman, R., & Siegle, E. (2015). Shoreline variability from days to decades: Results of long-term video imaging. *Journal of Geophysical Research: Oceans, 120*, 2159–2178. https://doi.org/10.1002/2014JC010329

Pontee, N. (2013). Defining coastal squeeze: A discussion. *Ocean & Coastal Management, 84*, 204–207. https://doi.org/10.1016/j.ocecoaman.2013.07.010

Ponting, J., & Mach, L. (2025). Big fines and fake signs: Lessons for natural recreation resource governance from surfing in the time of COVID-19. *Sport in Society, 28*(2), 234–257. https://doi.org/10.1080/17430437.2025.2457821

Pucino, N., Kennedy, D. M., Carvalho, R. C., Allan, B., & Ierodiaconou, D. (2021). Citizen science for monitoring seasonal-Scale Beach erosion and behaviour with aerial drones. *Scientific Reports, 11*(3935). https://doi.org/10.1038/s41598-021-83477-6

Reiblich, J., & Reineman, D. (2018). Rhino chasers and rifles: Surfing under the public trust doctrine. *Journal of Land Use & Environmental Law, 34*(1) https://ssrn.com/abstract=3382535

Reineman, D., Thomas, L. N., & Caldwell, M. (2017). Using local knowledge to Project Sea level rise impacts on wave resources in California. *Ocean & Coastal Management, 138*, 181–191. https://doi.org/10.1016/j.ocecoaman.2017.01.020

Rhode Island Coastal Resources Management Council. (2016). *RI sea grant launches new web-based public access app featuring CRMC rights-of-way.* http://www.crmc.ri.gov/news/2016_0822_rowapp.html.

R.I. Gen. Laws § 46-23-1(a)(1).

R.I. Gen. Laws § 46-23-26.

Ribot, J. C., & Peluso, N. L. (2003). A theory of access. *Rural Sociology, 68*(2), 153–181. https://doi.org/10.1111/j.1549-0831.2003.tb00133.x

Sandbrook, C. (2015). The social implications of using drones for biodiversity conservation. *Ambio, 44*, 636–647. https://doi.org/10.1007/s13280-015-0714-0

Setter, R. O., Han, R. X., Tavares, K.-D., Newfield, C., Terry, A., Roberson, I. M., Tarui, N., & Coffman, M. (2023). Managing retreat for Sandy Beach areas under sea level rise. *Scientific Reports, 13*(11920). https://doi.org/10.1038/s41598-023-38939-4

State v. Ibbison, 448 A.2d 728 (R.I. 1982).

Sweet, W. V., Hamlington, B. D., Kopp, R. E., Weaver, C. P., Barnard, P. L., Bekaert, D., Brooks, W., Craghan, M., Dusek, G., Frederikse, T., et al. (2022). Global and Regional Sea level rise scenarios for the United States: Updated mean projections and extreme water level probabilities along U.S. coastlines. NOAA Technical Report NOS 01 (p. 111). National Oceanic and Atmospheric Administration, National Ocean Service, Silver Spring, MD. https://oceanservice.noaa.gov/hazards/sealevelrise/noaa-nostechrpt01-global-regional-SLR-scenarios-US.pdf.

Tex. Nat. Res. Code Ann. § 61.011 (Vernon).

Thompson, R. (2015). Local government and the closing of the coast: Parking bans and the beach as a traditional public forum. *Fordham Environmental Law Review, 25*(2), 458–508. https://ir.lawnet.fordham.edu/elr/vol25/iss2/6

Titus, J. G. (2011). *Rolling easements, climate ready estuaries program.* Environmental Protection Agency. http://water.epa.gov/type/oceb/cre/upload/rollingeasementsprimer.pdf.

U.S. Department of Transportation. General Counsel. (2023). In J. E. Putnam, & M. A. Nichols. *Updated fact sheet (2023) on state and local regulation of Unmanned Aircraft Systems (UAS).* Washington, D.C. https://www.faa.gov/sites/faa.gov/files/State-Local-Regulation-of-Unmanned-Aircraft-Systems-Fact-Sheet.pdf.

Virginia Institute of Marine Science. '*Catch the king' tide.* Effective November 23, 2024. https://www.vims.edu/people/loftis_jd/Catch%20the%20King/.

Young, S. S., Rao, S., & Dorey, K. (2021). Monitoring the erosion and accretion of a human-built living shoreline with drone technology. *Environmental Challenges, 5*(100383). https://doi.org/10.1016/j.envc.2021.100383.

Chapter 13
Why the Printed Surf Mag Still Matters

Craig P. Sims

Introduction

A sport subculture is defined by Jary and Jary (2006) as, "any system of beliefs, values and norms which is shared and actively participated in by an appreciable minority of people within a particular culture" (in Ford & Brown, 2006, p. 59). Stranger (2011) explains that the surfing subculture is grounded by the act of riding waves—which he describes as the "pursuit of an ecstatic communion with nature" (p. 118). Based on this, the surfing subculture can be defined as a set of beliefs, values and norms grounded by the act of riding waves in an ecstatic communion with nature. Olive et al. (2015) explain that subcultural norms and values are constantly evolving as they are negotiated daily by the people and contexts within it. Despite this she suggests that the surf media can be, "as productive of a culture as they are reflective of it" (p. 262). Stranger (2011) suggests the surf media's influence is more direct, explaining that they play a significant role in the development and maintenance of "a global collective conscience" by presenting a style guide from which both surfers and non-surfers draw inspiration (Stranger, 2011, p. 144). Surf magazines have been integral to this process, having played a seminal role in creating and disseminating the sport's subcultural values since the 1960s (Booth, 2008; Ford, 2005; Stranger, 2011).

The aspirational appeal of surfing's subcultural values attracts consumers of surfing-related goods and services and many of these consumers are not active participants in the sport of surfing. Indeed, it is widely reported that surfing's consumer base is substantially larger than its participant base (Jarratt, 2010; Ford & Brown, 2006; Sims, 2022; Stranger, 2011). Given the centrality of surf media, particularly magazines, to the creation, maintenance, and dissemination of these subcultural

C. P. Sims (✉)
Bond University, Gold Coast, QLD, Australia
e-mail: csims@bond.edu.au

L. Mach (ed.), *The Future of Surfing in the Anthropocene*,
https://doi.org/10.1007/978-3-032-09377-6_13

values, their role in supporting the underlying drivers of demand for producers of surfing-related goods and services is profound.

However, the technological progression of media has changed the way consumers value the utility of print media (Bonner, 2015; Sims, 2022; Webb & Fulton, 2019), spearheaded by what is perceived to be a wholesale abandonment of traditional print media by youth, in favour of digital media (Bonner & Roberts, 2017). Surf magazines have not been excluded. The most notable examples were the closure of two pillars of global surf culture—America-based Surfing (2017) and Surfer (2020) magazines. These closures sent shockwaves through the surfing world and ignited fresh debate about the future of magazines in the surfing world (Warshaw, 2017; Doherty, 2020). Indeed, Warshaw (2017) declared the closure of Surfing Magazine after 53 years in publication as "ending the reign of a goliath in surf media and signalling the imminent and cold, 'Print Is Dead' future" (para. 1).

Nearly a decade later, the surf magazine sector has shown remarkable resilience with many still in circulation. As recently as August 2024, Surfer Magazine relaunched as a regular print product receiving widespread approval from surfers around the world. A contributing factor, and one that has been widely referenced in the scholarly literature, concerns declining levels of trust in social media and open digital content platforms (Edelman, 2023). The rise and spread of misinformation and disinformation, also known as *fake news*, has led Edelman (2018) to describe the resulting low levels of media trust as "the existential challenge of our times" (p. 3). In a media landscape dominated by digital content creation and delivery, the stubborn resilience of the printed surf mag and its continued influence as a driver of demand, begs some investigation.

The demand for surfing-related goods and services is met by the surf industry which is defined as a collective of commercial enterprises that have developed "lucrative niche markets" (Ford & Brown, 2006, p. 56) by successfully articulating surfing's cultural "values and symbolisms" (p. 57) to both surfers and the wider non-surfing society. The surf industry is one component of a wider interconnected network; the surf ecosystem, which is a broad term that encapsulates all the coexisting and interrelated components of surf culture. These components of surf culture include surfing participants, surfing consumers, the surf industry, and the surf media.

Applying a systems perspective which highlights the perennial existence of cause-and-effect relationships between seemingly disparate entities within a system, this chapter explores the role of the printed surf mag in the surfing ecosystem, so we have a better understanding of its potential purpose in the Anthropocene. After presenting a brief account of the evolution of surf media, the chapter reviews some of the theoretical approaches to scholarly work in this field. Then it details some of the empirical findings that support this chapter's assertion that despite digital disruption, the printed surf magazine is and is likely to remain an important part of the surfing ecosystem.

A Brief History of the Surf Media

In the early stages of surfing's revival, after having been nearly eliminated from Hawaiian culture by missionaries in the late nineteenth century, books on surfing played a significant role in shaping and expanding surf culture (Jarratt, 2010). Little research is available of the history of surf books, the earliest form of specialised surf media, but *The Surf Riders of Hawaii* (1914), self-published by Alfred Gurrey Jr., is believed to be the first book dedicated to surfing (SurferToday, n.d.). The book consisted of six leaves of paper containing eight photo prints accompanied by captions and poems that reflected the spirit of Hawaiian surfing. Over the decades that followed, hundreds of books were published focusing variously on surf history, biographies, literary works, technique and equipment guides, but most commonly, photography collections. An example being Ted Grambeau's self-published book, *Lost in Light*, containing over 200 images from a lifetime of surf travel. Printed in large format, in brilliant colour, with a hard cover embossed with his logo, it arrives with a sense of occasion in a custom-made box.

Despite its common connection to the printing press, the periodical surf magazine did not evolve from surf books. The lineage to magazines occurred, unusually, through film. During the 1950s, the surf film was the dominant medium that facilitated an exchange of ideas among surfers (Jarratt, 2010). Most came out of Southern California, where surfing's popularity was germinating. Bud Browne, a schoolteacher at the time, produced a film he named *Hawaiian Holiday (1954)*, conducting multiple screenings to packed audiences from Santa Monica to San Francisco. While not the first, and not Browne's first, *Hawaiian Holiday* marked the genesis of the commercial surf film industry (Jarratt, 2010). A few years later, prominent big wave surfer Greg Noll produced a series of movies, all called *Search for Surf* (1957), featuring exotic locations like Australia, Mexico and the Hawaiian island of Molokai, where surfing was still relatively unknown. Venturing out beyond home shores in search of the perfect wave is an alluring feature of surf culture, and evidently, it has been a focus of the surf media since its inception.

One of the surfers featured in Noll's films was Bruce Brown (no relation to Browne with an e), who later became a talented film maker and took the surf travelogue genre to a whole new level, with the release of the film *The Endless Summer* (1964). The idea was simple, when it is Winter in the Northern Hemisphere, then it is Summer in the Southern Hemisphere and so, by shifting hemispheres it is possible to live the perfect surfing life following the Summer endlessly. This became the storyline of *The Endless Summer;* two surfers Robert August and Mike Hynson travelling the world in a never-ending Summer of adventure and surf discovery. In 1966, after converting the film to a 35mm format suitable for mainstream cinemas, the film was subsumed into the Hollywood complex and screened at cinemas around the world, fuelling the popularity of this growing sport.

The most influential film of this era, however, was not one that emerged from the surfing community. *Gidget* (1959) was a Hollywood movie produced by Columbia Pictures about a girl named Gidget who spent the summer fraternising with surfers on

Surfrider Beach in Malibu. The story reflected the surfing lifestyle in all its "wine-swilling, wave-ripping, sex-crazed" glory occurring up and down the Californian coast (Jarratt, 2010, p. 64). The film was distributed world-wide showcasing a Hollywood take on the surfing lifestyle to a global audience. This activated dormant markets for surf cultures in places like Europe and South Africa, and accelerated the growth in existing markets like Australia, North America and Hawaii. While the influence of *Gidget* on surfing's popularity was profound, Bruce Brown laments most surfers were unhappy about the way their sport was depicted.

> The first Hollywood movies about surfing like Gidget and Beach Blanket Bingo gave the sport a bad rep. They made us out to be a bunch of idiots having food fights. We wanted to show how it really was: a legitimate sport. (Hoad, 2017, para. 1)

Attempts by those outside of the surfing subculture to depict the surfing way of life have always been met by criticism from surfers and so, while *Gidget* fuelled the public's fascination with surf culture, it was *The Endless Summer* that resonated with surfers.

A few months after *Gidget* the movie was first released, a young surfer and talented artist named John Severson, who dabbled in surf movies while doing military service in Hawaii, produced a 36-paged black and white magazine to promote his surf movie called *Surf Fever* (1960). He named the magazine *The Surfer* and sold it at his movie showings. He printed 5000 copies and it sold out before the movie tour had completed (Jarratt, 2010). In 1962, *Surfer* was launched as a quarterly magazine and would remain a pillar of surf culture for another six decades. While Severson has been touted as "the single most important voice in the creation of twentieth-century surf culture" (Ford & Brown, 2006, p. 28), *Surfer* magazine was not surfing's first dedicated magazine. That accolade belongs to the Californian publication *Reef*, which launched in August 1960 and lasted just three editions (Grisman, 2023). In fact, surfing featured prominently in magazines as far back as August 1888 when an illustration of a surfer was featured on the cover of the American periodical *The Police Gazette.*

Subsequently, surfing was depicted on the cover of various non endemic magazines 61 times prior to the release of *Reef* in 1960. For example, images of surfing graced the cover in mainland USA in *The Outing* (1908) and *Mid-Pacific* (1911) magazines, in Great Britain in *Scout* magazine (1932), in Hawaii in *Paradise of the Pacific* magazine (1932 and 1936) and in Australia's *Life* magazine and *The Queenslander Annual* (both in 1937). Most of these cover images were paintings or illustrations (see Fig. 13.1), and the articles inside them depicted surfing as a novel activity enjoyed by daredevils of the ocean, best observed from a safe distance, and usually with the same curiosity and delight as a tourist on safari watching wildlife in the African bushveld (A. Hunt, Personal Communication, October 27, 2024).

During the late 1960s and 1970s, books, magazines and film productions were driving the diffusion of surf culture around the world. Collectively, the surf media reflected the spirit of the surfing lifestyle which contained elements of freedom, adventure, hedonism and a distinct resistance to the mainstream values of the day. The

Fig. 13.1 Historic non endemic magazine covers

aspirational appeal of this lifestyle proposition attracted the interest of the non-surfing public, creating a new consumer base that fuelled the fledgling surf industry.

Through the late 1970s, 1980s and the early 1990s the surf media—dominated by magazines—and the surf industry were engaged in a productive and effective symbiosis with surf brands proving advertising support to the magazines that fuelled the subculture they profited from. Brands sought to express their importance in the subculture through their advertising placements. Surf magazine editors were gate-keeping powerbrokers whose editorial decisions had the capacity to affect both brands and the sponsorship value of their surfers.

In the late 1980s, the surf video joined the fray. Lower barriers to entry attracted budding surf film makers who churned out video productions, typically featuring high action surfing and high tempo music, in quick succession. Surf brands also embraced the technology, releasing video productions of their sponsored team riders which often become DVD cover mounts on surf magazines. Stranger (2011) summed up the pre-internet surf media landscape as follows:

> Collections of surfing magazines are pored over countless times and favourite images adorn the walls of surfers' bedrooms and beyond. Surf videos are watched repeatedly, either as whole programs or in fragmented snippets as viewers scan for their favourite sequences or watch short segments as the whim and opportunity coincide. (p. 188)

By the mid-1990s, the arrival of the Internet attracted copious investment along with the predicted death of any media without a domain name. After an initial splutter, known as the dot com crash, online-only media producers proliferated. Traditional media producers were slow to respond as they grappled to adjust their business models. Surf magazines, for example, hastily built websites and later added social media channels and ecommerce systems as extensions of their print-based business model but they were quickly overwhelmed.

At the turn of the century, low barriers to entry enabled anyone, and indeed everyone, to be a publisher and distributor of surf content. Independent, online-only media were joined by surf brands who invested in creating and disseminating their own media to promote themselves and their sponsored surfers. High profile sponsored surfers also realised that the internet, social media especially, provided a platform to connect directly to their fans. Within the space of just a few years, the surf media landscape became fractured and overcrowded. Surf magazines, once the gatekeepers to the culture—the influencers of the pre-internet era—were under threat.

Surf Media Scholarship

While there is a growing body of scholarly work exploring and explaining the many facets of surfing, very little research has been conducted into surf media itself and much of what has been studied concerns surfing's masculine hegemony and its disturbing track record of sexism and misogyny (Sims & O'Brien, 2025). Studies that have focused on surf media directly include—but are not limited to—Booth (1996) who categorised nearly five decades of the surf film and video industry into distinctive genres; Sims (2022) and Sims et al. (2025) who considered the future of the printed surf magazine in the digital age; and, Boyd (2024) who analysed the effect of Irish surf films on the commercialisation of Irish surf culture. The theoretical scope of studies into surf media are as broad and varied as the fields of inquiry, however, media substitution (MS) theory and uses and gratifications (U&G) theory are effective theoretical approaches that can help understand responses to the abundance of choice and ongoing disruption in the current media landscape (Sims & O'Brien, 2025).

MS theory is a suitable approach for examining the effect of digital disruption (Blumler & Katz, 1974) and asserts that when a new media technology, such as the Internet, is introduced, audiences redistribute the allocation of their time among available media options and as a result, new patterns of media consumption emerge (Kaye & Johnson, 2003). U&G theory explores how individuals actively make media choices for the gratification of their needs (Katz et al., 1973). By examining the motives behind the consumption of a particular form of media, researchers can understand the reasons for that medium's popularity, or failure, and the roles that the medium fills in society.

Applying MS theory while researching the early impact of digital media on traditional media, Adoni and Nossek (2001) advanced three possible outcomes relating to the extent to which one media option is substituted for another: (1) *Displacement*: as a result of the new medium being *functionally equivalent*, creating conditions that cause the incumbent to become either displaced or obsolete; (2) *Coexistence*: from a process of *functional differentiation* whereby the incumbent makes adjustments that facilitate a functional point of difference and re-establish its uniqueness; and, (3) *Convergence*: which is a *functional synthesis* of the two media sources resulting in a new medium that simultaneously utilises the functionality of both.

Relating the work of Adoni and Nossek (2001) to the surf media landscape, during the mid to late 1990s and beyond, many magazines were overwhelmed by the flood of online-only surf media. With declining advertising and copy sales and tighter production budgets, magazine staff found themselves spread thin trying to curate their magazines while attempting to populate their own digital media channels on an ever-shortening news cycle. The result was that magazine content diminished in depth and substance and was no longer distinguishable from what was available for free via the Internet. Magazines caught in this downward spiral eventually suffered *Displacement* because the new digital surf media were *functionally equivalent*, creating conditions that cause the incumbent to become obsolete.

Other magazines fighting the same downward spiral attempted *Convergence*, which is an effort to 'pivot' to a new model that simultaneously utilises the functionality of the incumbent and the disruptor. Here many magazines attempted to 'go digital', by simply migrating their print content along with their content building conventions to a digital distribution model anchored by their website. Most of these efforts at *functional synthesis* failed because their print-based content building systems that were deeply entrenched were too slow for the fast-paced news cycle.

Coexistence is an option few surf magazines have attempted. This option seeks to identify and capitalise on the factors that facilitate a functional point of difference between the incumbent and the disruptor in an attempt re-establish the incumbent's uniqueness. Sims (2022) investigated the factors that contributed to the possible *Coexistence* of surf magazines in the digital age.

From a participant group of Australian surfers between 15 and 60-years-of-age ($n = 1039$), Sims (2022) identified the most-preferred media source from a set of six U&G motives for media choice; namely, Information, Entertainment, Social Interaction, Identity Development, Habit, and Trust. He found websites enjoy a strong association with Information, Entertainment and Habit; magazines have a strong association with Trust and Identity Development; and, social media is strongly associated with Social Interaction and Habit. Interestingly, it was found that the Gen-Z segment (15–21 years-of-age) of the participant group engaged in high use of social media but had low trust in them and, paradoxically, they engaged in low use of magazines but had high trust in them. While low trust in social media was found to derive from learned experience, their trust in magazines was found to be predicated on something less pragmatic. Respondents assumed magazines are more trustworthy than social media based on an assumption that magazines take time, effort and consideration by multiple people to compile, and are thus, more trustworthy.

A by-product of the trust attributed to print is a sense of legitimacy and authority. This amounts to what Strömbäck et al. (2020), and Kalsnes and Krumsvik (2019) refer to as 'medium credibility'. They found that trust in a message is not just dependent upon who the sender is (source credibility) but can also be impacted by the medium through which it is delivered. Thus, a sense of credibility (Haniff, 2012; Kalsnes & Krumsvik, 2019; Strömbäck et al., 2020) and authority (Koivunen & Vuorelma, 2022) seems to also arise from the trust that is attributed to print.

Taking an MS theory perspective, Sims (2022) then argued that in the face of extreme disruption, if surf magazines are going to survive and continue as a seminal component of the sport's subculture and development, they should focus on the option of *Coexistence*. This form of adaptation requires a detailed analysis of the surf media ecosystem, and the surf magazine's role within it. At its most basic level, such analysis involves identifying, then capitalising on print's strengths, especially those strengths that counteract digital's weaknesses. This process could surface print's innate utility and become the foundation of the surf magazine's survival, and perhaps even, revival.

Since Sims (2022) found that magazines had a strong association with Trust and Identity, he concluded that surf magazines could *functionally differentiate* by capitalising on the authority and credibility that comes with trustworthiness and the influence that this authority and credibility has on identity development. To this end, he suggested that printed surf magazines should focus on providing high quality, well-researched content to maintain the trust that has been attributed to the medium.

The Effect of Trust on Identity, Authenticity and the Perception of Value

The superior trust afforded to surf magazines has implications for their value to the surf industry. Elliott and Wattanasuwan (2015) postulate a person's identity formation is a life-long affair that is actively created and constantly validated. Thompson (1995) describes this construction of the self as a symbolic project, during which a consumer seeks out symbolic materials, which are weaved into "a narrative of self-identity" (p. 120). Elliott and Wattanasuwan argue that "all voluntary consumption carries, either consciously or unconsciously, symbolic meanings; if the consumer has choices to consume, he or she will consume things that hold particular symbolic meanings" (p. 134). Consumption choices, therefore, are based not only on a product's utility but also on the contribution its symbolic meanings make to the construction of the consumer's identity. Thompson (1995) further explains that symbolic resources for the construction of self can be obtained through lived and/or mediated experiences, with mediated experiences resulting from mass-communication culture and the consumption of media products. With trust in digital media at an all-time low (Edelman, 2018), the printed surf magazine is distinguishable as a credible, authoritative, and trustworthy resource that reflects and informs the progression of authentic surf culture through its editorial and advertising content. This makes them well suited to both the construction and validation of an individual's identity as a surfer, and an effective driver of demand for surfing-related products.

To more fully interrogate its role as a trusted purveyor of authentic surf culture it is necessary to touch upon the concept of authenticity. In sociological literature, authenticity plays a key role in developing and displaying identity within a subculture. In this context, authenticity is referred to as "a set of qualities that people in a particular time and place have come to agree represent an ideal or exemplar" (Vannini & Williams, 2020, p. 3). Thus, the word 'authentic' has no inherent meaning because its meaning is based on how we perceive it as representative of a subcultural ideal. Moreover, because subcultural norms and values are negotiated daily through the interactions of their participants (Olive et al., 2015), authenticity is a created construct that is constantly evolving. With such a moving target, a trusted source becomes a valued source. Indeed, many studies have found that the materiality or physicality of printed books and magazines helps to conceptualise them as desirable, collectable and valuable objects (Bonner, 2015; Nossek et al., 2015; Webb & Fulton, 2019). The characterisation of printed media as coveted physical objects should not be underestimated because physicality is something that cannot be translated through digital media platforms. The ephemeral nature of digital content also makes the temporality of print a significant differentiator. A surf magazine is fundamentally a contained capsule that, through its trustworthy editorial and advertising content, offers a permanent snapshot of surf culture at the time of publication. To this extent, surf magazines can also be characterised as trusted cultural artifacts.

Media Choice and Content Preference

Another finding by Sims (2022) related to content suitability. The study revealed that respondents recognised surf magazines as being best suited to the delivery of content in two out of the five content categories presented, namely; Technique and Equipment, and Travel. Websites were found to be best suited to the consumption of content about Surf Conditions, a finding validated by the popularity of websites like Surfline and Swellnet, while social media was found to be well suited to Interviews and news relating to Competition and Competitive Surfers. Based on this finding, Sims postulated that surf magazines should position themselves as the most authoritative source of opinion, instruction and knowledge by drawing on acclaimed experts in the subjects of surfing technique and surfing equipment. This approach will perpetuate their medium credibility and satisfy their suitability for the delivery of content under the category Technique and Equipment. And finally, since it was found that the Travel content category was also well suited to the medium, surf magazines should also prioritise content about surf travel, adventure and discovery through high quality photography and storytelling. The desire to travel in search of the perfect wave is deeply embedded into the psyche of all surfers (Ford & Brown, 2006; Ponting & O'Brien, 2015; Preston-Whyte, 2001; Scheibel, 1995) so this content category has broad appeal and represents a big opportunity for surf magazines to remain relevant and in demand, especially since so much of the spirit of travel is expressed through imagery.

The Primacy of the Still Surf Photograph

Photographic images are important part of surf culture. Indeed, Ford and Brown (2006) assert "the history and significations of surfing are overwhelmingly expressed through pictures of surfing heroes" (p. 42). These pictures of surfing heroes do more than capture a moment in time for posterity. They also evoke "a powerful aesthetic and contemplative experience" (p. 42) for the viewer. Ford and Brown (2006) explain that a surfer engages in an instinctive, reactive response as they traverse a breaking wave which requires a presence of mind that, "denies that momentary reflection which allows the possibility of imprinting an image upon the mind's eye" (p. 33). The observer of surfing sees it from a similar perspective; a flowing series of idiosyncratic responses by the surfer to the wave's changing form, which is over in seconds. The act of surfing on a wave is therefore an ephemeral event. Once completed, there is no trace of its existence. A surf photograph freezes a moment in that complex interaction between the surfer and the wave, allowing the viewer to take in all aspects captured within the frame—the sunlight, texture of the water, shape of the wave, surfer's posture, the position of the board, and the shards of spray emanating from it, as well as features of the background and the foreground.

A surf photo also offers what Barrett (1985) describes as 'interpretive ambiguity'. We identify detail that aligns with our level of ability, sense of place, and kinship with the culture so the memories and feelings it evokes are unique to each of us. To this extent, the still surf photograph represents not just the capture of reality but also the creation of it (Stedman, 1997). Revered surfing legend, Gerry Lopez expresses this as an awakening experience.

> Somebody once said, 'Nature speaks to you in a voice that's familiar to your soul'. In a surf magazine you have beautiful images - of something very natural - and they speak to your soul. A lot of times you don't even know it, but a picture will awaken something deeper than what's in your mind. It activates your senses, your feelings, and your imagination. (G. Lopez, Personal communication, October 10, 2022)

The transformative power of the still surf photograph is explored in Fig. 13.2. This image allows the casual viewer to appreciate the torque of the surfer's body, the position of the board on the wave and the spray emanating from it. The viewer might also notice the colours and textures of various parts the wave and the beauty of the natural setting in which the action is unfolding. An experienced surfer understands that the *down carve* is a highly technical manoeuvre that requires subtle adjustment in weight and body position as the surfer executes the turn. They might notice the twist in the rider's upper body with arms, shoulders and head directing the trajectory of the board while the legs and hips remain fixed, exerting pressure on the board as it progresses through the turn. Locals, recognising the headland in the background and the surfer as a familiar face in the lineup, might feel a sense of pride as their local spot is immortalised in print (or concern that it has now been exposed). On a closer or subsequent inspection, the viewer might pick up on the shadow of the surfer's shoulder, arm and fingers cast on the wave face near the surfer's right knee or the effect track of the surfboard's path has had of the cascading lip. To most surfers, the image will evoke a familiar sensation of carving down the face of a wave as the image invokes memories of their own similar experiences. Indeed, some will imagine the feeling of executing such a manoeuvre with such skill and be inspired to attempt it in future surf sessions. That this much can be gleaned from a single image speaks to the value of time spent on interpreting the detail in a surfing image. In the ephemeral, fast-paced digital media world, this is not typically the way content is consumed.

Fig. 13.2 The transformative power of the still surf photograph, Photography by Andrew Shield (www.andrewshield.com.au)

Figure 13.3 touches on both the power of the frozen moment and the allure of travel. In the fleeting passage of real time (if seen live or on video), the impact of this scene might be only lightly felt but with time spent exploring the detail, the various perspectives contained within the image can be appreciated. All surfers yearn to experience uncrowded perfection. Indeed, they actively seek it. At times this leads them to paddle to the next peak down the beach, drive around the next headland, or as this image depicts, travel to a distant and exotic destination. To the active surfer this image represents the ultimate reward to those who seek. The surfer getting tubed in front of his mates is being directly rewarded, while the surfer paddling out gleefully watches on knowing that he will be next. The surfer on the boat is probably about to experience a rush of adrenaline that will see him scurry for his board and hop into the water within a matter of seconds. These are all familiar emotions that are stirred when this image is digested by the active surfer. To the casual non-surfing viewer this image captures the adventurous spirit of the surfing lifestyle and while most would never find themselves in this position, it would not take much effort to imagine themselves living this life. This image fuels the aspirational appeal of surf culture and has the interpretive ambiguity to allow the viewer to contemplate how they would feel and what they would do if they were in this place at this time. Additionally, with every revisit this image has the potential to conjure different feelings depending on what captures their attention at the time. In the ephemeral, fast-paced digital media world, this image is unlikely to evoke such a contemplative experience.

Fig. 13.3 The power of the frozen moment and the allure of travel, Photography by Andrew Shield (www.andrewshield.com.au)

Conclusion

Narrative theory examines how the distinctive nature of stories and their various structures, elements, uses and effects shape and inform our perception of the world we live in. The surf media has over the passage of time, documented, showcased, and interrogated all aspects of the sport and culture and as such, they present an over-arching meta-narrative by which surf culture can reflexively make sense of and to itself (Ford & Brown, 2006).

Today, the surf media is a vibrant and dynamic collective that functions across all channels including books, magazines, films, websites, podcasts, and social media. Once a major feature of the surfing media landscape, printed surf magazines have struggled to navigate their way through digital disruption. It has been argued that by identifying and capitalising on their unique and distinctive features, printed surf magazines have the potential to achieve *Coexistence* (Adoni & Nossek, 2001) and remain relevant, influential, and in demand.

Due to the trust attributed to the medium, it has been shown how the printed surf magazine is an effective contributor to the maintenance and dissemination of surfing's cultural norms and values, providing ways in which surfers and surfing identify and distinguish themselves. And how, through this role, surf magazines are instrumental in driving demand for surfing related products by consumers whose identity is tethered to surf culture. It has also been shown how, through their superiority for showcasing the still surf photograph, printed surf magazines are particularly eloquent and resonant in their depiction of the surfing experience. And the physicality and temporality of surf magazines position them as important cultural artifacts that are valued for their authentic portrayal of the surfing lifestyle.

With these unique and distinctive features in mind, and looking to the future, surf magazine publishers and editors should focus on protecting, embedding, and leveraging the trust that is attributed to magazines. Chiefly, this can be achieved by delivering and promoting content that depends on its trustworthiness for impact, and to use the most trusted sources to originate this content. For example, tips about technique, reviews about equipment, and explanations of surf-related phenomena delivered by widely recognised experts in each field will reinforce the magazine's role as a trusted and authoritative source within a media landscape awash with misinformation, disinformation, and "fake news". Publishers and editors should also prioritise features and high-quality imagery about exotic and distant surfing locations. This is a content category that resonates with all surfers and the accompanying photographs have the capacity to temporarily transport readers from their mundane day-to-day lives and speak to them in a voice that's familiar to their soul (Lopez, 2022).

By being cognisant of the inherent strengths of the medium and focussing on editorial strategies that play to these strengths, the humble surf magazine can remain resilient in the face of digital disruption and continue to function as an important component of surfing's subculture and development well into the future.

References

Adoni, H., & Nossek, H. (2001). The new media consumers: Media convergence and the displacement effect. *Communications. The European Journal of Communication Research, 26*(1), 59–83.

Barrett, T. (1985). Photographs and contexts. *The Journal of Aesthetic Education, 19*(3), 51–64. https://doi.org/10.2307/3332643

Bonner, E. M. (2015). *Millennials and the future of magazines: How the generation of digital natives will determine whether print magazines survive* [Doctoral dissertation, The University of Alabama]. Retrieved from ProQuest (1600862).

Bonner, E., & Roberts, C. (2017). Millennials and the future of magazines: How the generation of digital natives will determine whether print magazines survive. *Journal of Magazine & New Media Research, 17*(2), 1–13.

Booth, D. (1996). Surfing films and videos: adolescent fun, alternative lifestyle, adventure industry. *Journal of Sport History, 23*(3), 313–327.

Booth, D. (2008). (Re)reading the surfers' bible: The affects of tracks. *Continuum (Mount Lawley, W.A.), 22*(1), 17–35.

Blumler, J. G., & Katz, E. (1974). The uses of mass communications: Current perspectives on gratifications research. *Sage Annual Reviews of Communication Research* (Vol. III). Sage Publications, Inc.

Boyd, S. (2024). Beyond the noise: The cultural (or subcultural) politics of Irish surf films. *Sport in Society, 27*(6), 946–964. https://doi.org/10.1080/17430437.2024.2334599

Brown, B. (Director). (1966). *The endless summer* [Film]. Cinema V distribution.

Browne, B., (Director). (1953). *Hawaiian surfing movie* [Film]. Bud Browne films.

Doherty, S. (2020, October 8). *Surfer magazine is no longer: 1960–2020.* Surfline. https://www.surfline.com/surf-news/surfer-magazine-no-longer-1960-2020/99454

Edelman. (2018). *2018 edelman trust barometer*. Edelman. https://www.edelman.com/sites/default/files/2018-06/2018_Trust_Barometer_Brands_Social_Media_Special_Full_Report.pdf

Edelman. (2023). *2023 Edelman trust barometer*. Edelman. https://www.edelman.com/sites/g/files/aatuss191/files/2023-03/2023%20Edelman%20Trust%20Barometer%20Global%20Report%20FINAL.pdf

Elliott, R., & Wattanasuwan, K. (2015). Brands as symbolic resources for the construction of identity. *International Journal of Advertising, 17*(2), 131–144.

Feldman, F. (2010). *What is this thing called happiness?* Oxford University Press.

Ford, N. J. (2005). *Surfing and social theory*. Taylor & Francis Ltd.

Ford, N., & Brown, D. (2006). *Surfing and social theory experience, embodiment and narrative of the dream glide*. Routledge.

Grisman, K. (2023). *Welcome to all surf magazines*. https://allsurfmagazines.com/

Haniff, Z. (2012). *Niche theory in new media: Is digital overtaking the print magazine industry?* [Masters thesis, University of Navada]. ProQuest Dissertations Publishing.

Hoad, P. (2017. December 14). *'If there's an ocean, maybe there's surf': Bruce brown on making the endless summer*. The Guardian. https://www.theguardian.com/film/2017/dec/13/how-we-made-the-endless-summer-bruce-brown-interview

Jarratt, P. (2010). *Salts and suits*. Hardie Grant Publishing.

Jary, D., & Jary, J. (2006). Collins web-linked dictionary of sociology (1st Collins U.S. ed.). Collins

Kalsnes, B., & Krumsvik, A. H. (2019). Building trust: Media executives' perceptions of readers' trust. *Journal of Media Business Studies, 16*(4), 295–306.

Katz, E., Blumler, J. G., & Gurevitch, M. (1973). Uses and gratifications research. *Public Opinion Quarterly, 37*(4), 509–523.

Kaye, B. K., & Johnson, T. J. (2003). From here to obscurity?: Media substitution theory and traditional media in an on-line world. *Journal of the American Society for Information Science and Technology, 51*(3), 260–273.

Koivunen, A., & Vuorelma, J. (2022). Trust and authority in the age of mediatised politics. *European Journal of Communication, 37*(4), 393–408.
Nossek, A. H., & Nimrod, G. (2015). Is print really dying? The state of print media use in Europe. *International Journal of Communication (Online),* 365–385. https://ijoc.org/index.php/ijoc/article/view/3549. Made available under CC BY-NC-ND 4.0.
Olive, R., McCuaig, L., & Phillips, M. G. (2015). Women's recreational surfing: A patronising experience. *Sport, Education and Society, 20*(2), 258–276.
Ponting, J., & O'Brien, D. (2015). Regulating "Nirvana": Sustainable surf tourism in a climate of increasing regulation. *Sport Management Review, 18*(1), 99–110.
Preston-Whyte, R. (2001). Constructed leisure space: The seaside at Durban. *Annals of Tourism Research, 28*(3), 581–596.
Randle, Q. (2001). *Gratification niches of monthly print magazines and the World Wide Web.* [Doctoral thesis]. Michigan State University. ProQuest Dissertations Publishing.
Rosenblatt, L. (1986). The aesthetic transaction. *The Journal of Aesthetic Education, 20*(4), 122–128.
Sax, D. (2016). *The revenge of analog: Real things and why they matter*. PublicAffairs.
Scheibel, D. (1995). "Making waves" with burke: Surf Nazi culture and the rhetoric of localism. *Western Journal of Communication, 59*(4), 253–269.
Sims, C. (2022). The nature of Gen-Z's influence on the future of printed surf magazines. Bond University.
Sims, C., O'Brien, D., Gowthorp, L., & Scott, O. (2025). The functional differentiation of the printed surf magazine in the digital age of leisure reading. Leisure Studies, 1–19.
Sims, C. P., & O'Brien, D. A. (2025). The Surf Industry. In Kennedy, D. (Ed.), *The science and culture of surfing* (Ch. 12).
Stranger, M. (1999). The aesthetics of risk: A study of surfing. *International Review for the Sociology of Sport, 34*(3), 265–276.
Stranger, M. (2010). Surface and substructure: Beneath surfing's commodified surface. *Sport in Society, 13*(7–8), 1117–1134.
Stranger, M. (2011). *Surfing life: Surface, substructure and the commodification of the sublime.* Ashgate.
Stedman, L. (1997). From Gidget to Gonad man: Surfers, feminists and postmodernisation. *Australian and New Zealand Journal of Sociology, 33*(1), 75–90.
Strömbäck, J., Tsfati, Y., Boomgaarden, H., Damstra, A., Lindgren, E., Vliegenthart, R., & Lindholm, T. (2020). News media trust and its impact on media use: Toward a framework for future research. *Annals of the International Communication Association, 44*(2), 139–156.
SurferToday. (n.d.). *'The Surf Riders of Hawaii': The story of the world's first surf book.* SurferToday. https://www.surfertoday.com/surfing/the-surf-riders-of-hawaii-the-worlds-first-surf-book#:~:text=%22The%20Surf%20Riders%20of%20Hawaii,copies%20are%20known%20to%20exist
Thompson, J. (1995). The media and modernity: a social theory of the media. Cambridge: Polity Press.
Vannini, P., & Williams, J. P. (2020). *Authenticity in culture, self, and society.* Routledge, Taylor & Francis Group.
Warshaw, M. (2017, May 6). Rip surfing magazine (1964–2017). Surfline. https://www.surfline.com/surf-news/rip-surfing-magazine-1964-2017/1408
Webb, S., & Fulton, J. (2019). "I want to read it in my hands": The aesthetic attraction of independent women's magazines. *Australian Journalism Review, 41*(2), 273–287.
Wheaton. (2004). *Understanding lifestyle sports: Consumption, identity, and difference.* Routledge.

MIX
Papier aus verantwortungsvollen Quellen
Paper from responsible sources
FSC® C105338

If you have any concerns about our products,
you can contact us on
ProductSafety@springernature.com

In case Publisher is established outside the EU,
the EU authorized representative is:
Springer Nature Customer Service Center GmbH
Europaplatz 3, 69115 Heidelberg, Germany

Printed by Libri Plureos GmbH
in Hamburg, Germany